RS•C

Foreword

Chemistry is a conceptual subject and, in order to explain many of these concepts, models are used to describe and explain the microscopic world and relate it to the microscopic properties of matter.

As students progress in chemistry the models they use change and many contradict their everyday experiences and use of language. This resource is designed to provide some strategies for dealing with some of the misconceptions that students have.

It is hoped that teachers will see the benefit of a constructivist approach and find that the student material leads to a good theoretical underpinning of the fundamentals of chemistry.

Professor Steven Ley CChem FRSC FRS
President, The Royal Society of Chemistry

RS•C

Contents

Probes in group A are most suitable for the 11–14 age range.
Probes in group C are most suitable for post-16 courses.
Probes in group B can be used at 14–16 level or for post-16 courses.

RS•C

Acknowledgements

I would like to acknowledge the cooperation of a large number of people and organisations who helped with this project in various ways. The following colleagues and institutions trialled or commented on draft materials, or sent useful suggestions or information. I hope I have remembered everyone, but with such a collective effort I trust any omissions will be forgiven.

Glen Aikenhead – University of Saskatchewan, Canada
Jon Angell – Downend School
Stuart Barker – Park College
Michael Barton
Richard Biddle – Carmel Technology College
Andrew Biggs – Shrewsbury School
Peter Biggs – Hinchinbrooke School
Vanessa Bird – Worth School
Marc Bloch – Heinemann Educational Publishers
John Bloor – Universities of Virginia and Tennessee, USA
Martin Bluemel – Taunton School
Anne Brearley – United World College of the Adriatic, Italy
Peter Buck – University of Education, Heidelberg, Germany
Mike Clugston – Tonbridge School
David Cooper – Sutton Valence School
Di Cook – Wispers School
Andrew Davies – St Boniface's College, Plymouth
Gaynor Davies – Ounsdale High School
Philip Dobson – Berkhamsted Collegiate School
Michael Dorra – University of Education, Heidelberg, Germany
Keith Fleming – University of New England, Australia
Duncan Fortune – Glasgow Caledonian University
Mike Foster – St Margaret's School, Exeter
Mark Gale – South Dartmor Community College
John Gilbert – University of Reading
Grant Gill – Kemnal Technology College
Alistair Gittner – King's School, Peterborough
Alan Goodwin – Manchester Metropolitan University
Richard Grime – The King's School, Macclesfield
Anthony Hardwicke – Monmouth School
Claus Hibling – Institute for the Teaching of Chemistry, Münster, Germany
Geeske van Hoeve-Brouwer – Eindhoven University of Technology, The Netherlands
Andrew Hunt – Nuffield Curriculum Projects
Anne Hurworth – Park Lane College, Leeds
June Jelly – Menzieshill High School
Bill Johnson – International School of Geneva
Philip Johnson – University of Durham
John Kerr – Winchester College
Ros Key-Pugh – Royal High School
K. G. King – Perse School for Girls
Sarah Knight – James Allen's Girls' School
Gill Kuzniar – Netherhall School
Michael Laing – University of Natal
Helen Langslow – Stamford High School
Mark Leach
Christine Lewis – Keswick School

RS•C

Steve Lewis – Shrewsbury Sixth Form College
John Luton – Varndean College, Brighton
Robin Millar – University of York
Roger Mitchell
Raphael Mordi – Newcastle College
Matthew Morrison – Strathcona Baptist Girls' Grammar School, Australia
Brian Murphy – United Arab Emirates University
Jim Murphy – St. Boniface's College, Plymouth.
Paul Murphy – Loretto School
Vincent Murphy – Copenhagen International School
David Neill – Canford School, Wimborne
Igor Novak – National University of Singapore
Jerry O'Brien – Slough Grammar
Philip O'Connor – Oakwood Park Grammar School
John Oversby – University of Reading
Andrew Page – Braintree College
Lambros Papalambros – Livadia High School, Greece
Nicole Pearce – Headlands School
M. Arminda Pedrosa – University of Coimbra, Portugal
Margaret Price – Balfron High School
K. C. Pun – Tsang Shiu Tim Secondary School, Hong Kong
Alan Quinn
David Reynolds – Long Road Sixth Form College, Cambridge
Graham Riley – King Edward VI Grammar School, Chelmsford
Keith Ross – Cheltenham and Gloucester College of Higher Education
Wendy Rudge – Canterbury College
Charly Ryan – King Alfred's College, Winchester
Eric Scerri – University College of Los Angeles
Hans-Jürgen Schmidt – University of Dortmund, Germany
Colin Smith
Mrs. S. Smith – Hazelwick School
Marinella Spezziga
Lesley Stanbury – St. Albans School
Julia de Ste Croix – East Devon College of Further Education
Ann-Marie Stott – St Paul's Catholic College, Haywards Heath
Ben Styles – University of Sussex
Joanne Sumner – Maricourt High School
Mohammed Taj – Deacons School
Daniel Tan – Nanyang Technological University, Singapore
Zoë Thorn – Saffron Walden County High School
Tony Tooth – King's School, Ely
Georgios Tsaparlis – University of Ioannina, Greece
Zoltán Tóth – University of Debrecen, Hungary
Nicos Valanides – University of Cyprus
Christina Valanidou – University of Cyprus
David Waistnidge – King Edward VI College, Totnes
Andrew Wallace – Durham School
Paul Warren – Linton Village College
Simon Warburton – Ferrers School
Mike Watts – University of Surrey Roehampton
Jonathan Wilkinson – Bingley Grammar School
Bob Wright – Charterhouse School

RS•C

I would also like to thank all the correspondents on the learning-science-concepts e-mail discussion list (**http://www.egroups.co.uk/group/learning-science-concepts**) for some stimulating and provocative exchanges about some of the issues discussed in this publication.

Most of the classroom materials presented in this resource pack were written by the author, although informed by published or other research. The probe on Mass and dissolving was based on a question written by Dr. Vanessa Barker of the Institute of Education, University of London. The probes on Elements, compounds and mixtures were based on a question set as part of a National Survey undertaken by the Asessment of Performance Unit of the (then) Department of Education and Science in the U.K. The other classroom materials are original, but informed by a variety of existing research which is cited in the text.

Above all, thanks are due to the Royal Society of Chemistry for funding the project and awarding the Teacher Fellowship; Homerton College (and in particular, the Principal, Dr Kate Pretty) and the Faculty of Education, University of Cambridge for releasing me from teaching duties; The University of London Insitute of Education for awarding a Visiting Fellowship, and the Science & Technology Group there for providing a welcome, comradeship and logistical support; Dr Colin Osborne (RSC Education Manager, Schools and Colleges) and Dr Maria Pack (RSC Assistant Education Manager, Schools and Colleges) for support and for trusting my instincts; members of the RSC Committee for Schools and Colleges for commenting on drafts; and Philippa Taber for some clerical assistance with data analysis, but mostly for generous tolerance of my obsessions.

Keith Taber
August 2001

RS•C

How to use this resource

Electronic versions of the worksheets are available at
www.chemsoc.org/networks/learnnet/miscon2.htm which may be downloaded and customised.

[*sic*] – this word is used to indicate an apparent mispelling or doubtful word or phrase in the source
being quoted.

Throughout this resource the referral to 'Teachers' notes' indicates the companion Theoretical
background volume to this publication. Links between the two volumes are summarised below:

RS•C

Teacher's feedback form

Chemical misconceptions – prevention, diagnosis and cure Volume II Classroom Resources

The Royal Society of Chemistry welcomes feedback on how useful teachers find this publication.

If you have any comments on the book, or on specific resources or suggestions included in it, please make a copy of this page and send your comments to

Dr. Keith Taber
c/o Education Department
The Royal Society of Chemistry
Burlington House
Piccadilly
London W1J 0BA

Please include an email address if you would like an acknowledgement of, or response to, your comments.

Teacher's name:

_____

Institution name:

_____

Email address (optional):

_____

Your comments:

_____

_____

_____

_____

_____

_____

_____

_____

_____

_____

_____

_____

_____

_____

_____

Thank you for taking the time to let us know your views on these materials.

RS•C

This page has been intentionally left blank.

RS•C

Elements, compounds and mixtures

Target level

These materials are primarily intended for the 11–14 age range, but may also be used as revision materials for the 14–16 age range.

Topics

Pure substances and mixtures; elements and compounds.

Rationale

The distinctions between pure substances and mixtures, and between elements and compounds are fundamental in chemistry. The materials comprise a pre-test, a study activity, and a post-test. The study activity is an exercise that focuses first on the distinction between pure substances and mixtures (one or several types of molecule present), and then distinguishes between single substances that are elements and compounds (one or more type of atomic core or nucleus or atom present in the molecule). These ideas are discussed in Chapter 6 of the Teachers' notes.

During piloting, some teachers found the materials 'very useful', 'very revealing', 'very clear indeed' and 'helpful'. The pre-test revealed 'considerable uncertainty' about definitions. It was also found that some students had difficulty accepting the more complex molecules represented (such as a benzene molecule), and one teacher found that students 'were reluctant to relinquish their own ideas' (the nature of learners' alternative conceptions is discussed in Chapter 1 of the Teachers' notes). Some students saw bonds as concrete structures and expected them to be drawn as solid lines (see Chapter 6).

Teachers reported 'improved understanding' when students used the materials. One teacher though that student responses were 'considerably more coherent after the exercise', and another commented that 'weaker pupils began to sort out the difference'. Some students reported that they enjoyed the activity, and that it helped them understand what the key terms meant.

Several teachers did not like the introduction of the word 'core' and could substitute 'nucleus' or 'atom' if preferred. Some students in the pilot objected to monatomic molecules being described as molecules (see Chapter 2 of the Teachers' notes for a discussion of the definition of 'molecule'). The exercise was considered too lengthy and repetitive for some students. The full version is provided here, but individual teachers may wish to edit it, to match the age and ability of particular classes.

Details of the DARTs activity can be found in Chapter 5 of the Teachers' notes.

Instructions

There are three sets of worksheets;

the pre-test: **Elements, compounds or mixtures? (1)**, the study task: **Elements, compounds and mixtures (2)**, and the post-test: **Elements, compounds or mixtures? (3)**.

RS•C

If the materials are used after teaching the topic, then the pre-test may be used to diagnose whether or not students will benefit from working through the study task.

Resources

■ Student worksheets
  – Elements, compounds or mixtures? (1) (pre-test)
  – Elements, compounds and mixtures (2) (study task)
  – Elements, compounds or mixtures? (3) (post-test)

Feedback for students

A suggested answer sheet for the use of teachers is provided.

RS•C

RS•C

Elements, compounds and mixtures – answers

Elements, compounds or mixtures? (1) (Pre-test)

Various definitions of elements and compounds may be used (see Chapter 2 of the Teachers' notes). The following is suggested. Atom or nucleus can be used instead of core.

1. An element is a pure substance which contains identical atoms or molecules with only one type of atomic core.

2. A compound is a pure substance which contains identical molecules with two or more types of atomic core.

3. A mixture is a material which has two or more types of molecules.

   Clearly these definitions only apply to molecular materials, and will need to be extended to ionic and metallic materials later.

4. Compound - one type of molecule; each molecule has more than one type of atomic core.

5. Element – one type of atom or molecule, with only one type of atomic core.

6. Element – one type of atom or molecule, with only one type of atomic core.

7. Mixture (of two compounds) – two types of molecule present.

8. Mixture (of two elements) – two types of atom present.

9. Compound - one type of molecule; each molecule has more than one type of atomic core.

Elements, compounds and mixtures (2) (Study task)

1. Single substance

2. Single substance

3. Mixture

4. Mixture ... molecule

5. Single substance ... molecule

6. Compound ... [atomic] core [or atom or nucleus] ...

7. Element ... [atomic] core [or atom or nucleus] ...

8. Compound

9. Mixture

10. Element

Elements, compounds or mixtures? (3) (Post-test)

1. Compound

2. Element

3. Mixture

4. Compound

5. Mixture

6. Element

Elements, compounds or mixtures? (1)

In science, it is important to know the difference between elements, compounds and mixtures. Try to explain what you think each of these words means:

1. An element is

_____

_____

_____

_____

2. A compound is

_____

_____

_____

_____

3. A mixture is

_____

_____

_____

_____

On the following sheets you will find six diagrams showing the particles in some samples of materials.
The different particles are shown as:

Each diagram is meant to show either an element, a compound or a mixture.

Decide whether each diagram represents an element, a compound, or a mixture, and explain your reasons.

RS•C

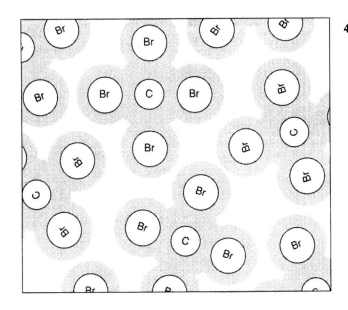

4. This diagram shows particles in

_____

I think this because

_____

_____

_____

_____

_____

_____

_____

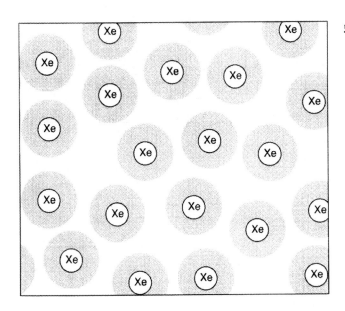

5. This diagram shows particles in

_____

I think this because

_____

_____

_____

_____

_____

_____

_____

_____

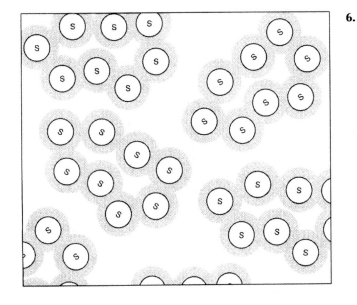

6. This diagram shows particles in

_____

I think this because

_____

_____

_____

_____

_____

_____

_____

_____

RS•C

Elements, compounds or mixtures? (1) – page 2 of 3

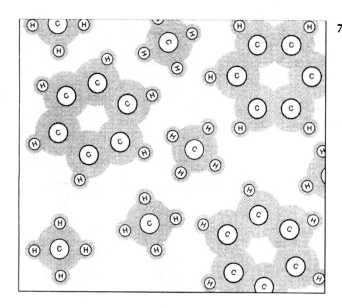

7. This diagram shows particles in

_____

I think this because

_____

_____

_____

_____

_____

_____

_____

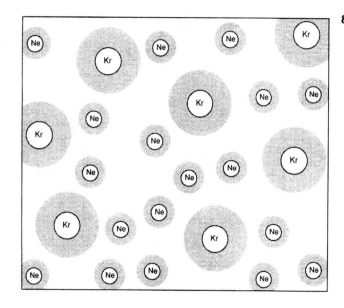

8. This diagram shows particles in

_____

I think this because

_____

_____

_____

_____

_____

_____

_____

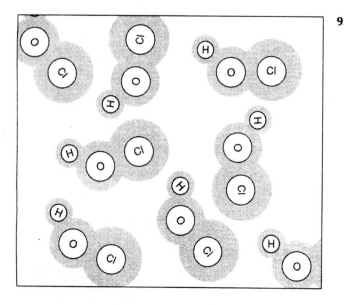

9. This diagram shows particles in

_____

I think this because

_____

_____

_____

_____

_____

_____

_____

_____

Elements, compounds or mixtures? (1) page 3 of 3

RS•C

Elements, compounds and mixtures (2)

Pure substances and mixtures

In science, it is important to know the difference between pure substances and mixtures of several substances. Scientists think about the differences in terms of the particles which are in the materials.

Scientists believe that all matter (all solids, liquids and gases) is made up from tiny particles that are much too small to be seen.

The tiniest particles are given names like 'electron' and 'proton'. These are arranged into slightly larger (but still very tiny) particles called atoms, ions and molecules. In many materials the particles are called molecules.

Here are some pictures that scientists use to represent atoms and molecules.

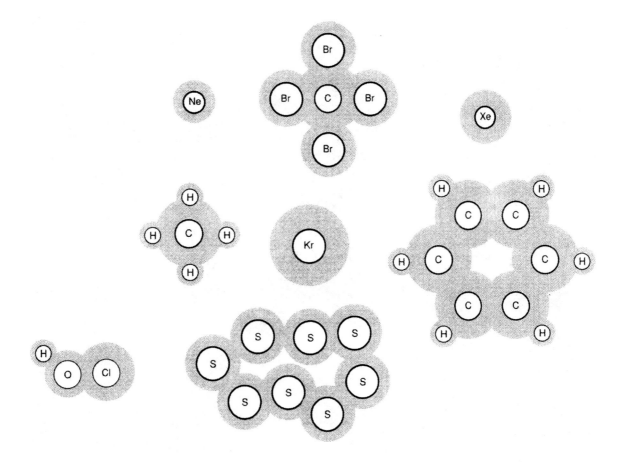

The letters are labels used by scientists to help identify the particles.

There are many different types of atoms and molecules, and these pictures just show a few examples.

Different substances contain different molecules

The three diagrams below show two different substances. Which two diagrams show the same substance?

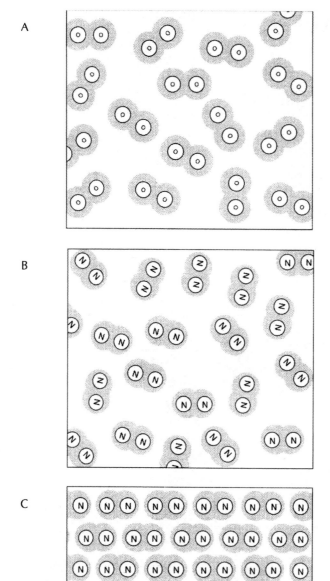

Answer

B and C are the same substance because they contain the same molecules. They look different because they show how the molecules are arranged when the substance is a gas (B) and a solid (C). Diagram A shows a different substance, because it shows different molecules.

Some materials only contain one type of molecule or atom. These are called single substances (or pure substances).

RS•C

The following diagrams show some single substances. They are each different substances because they have different types of molecules.

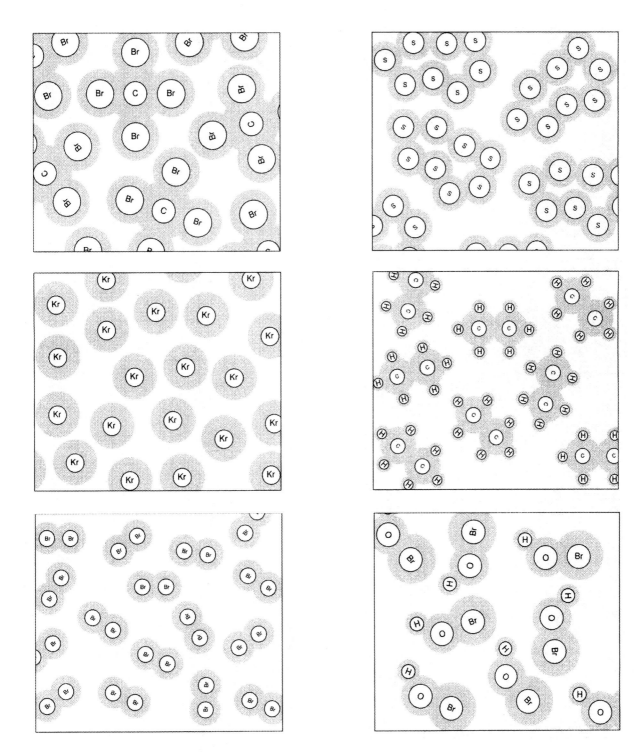

When substances are mixed, their molecules become mixed up. We call the new material a mixture.

A mixture contains more than one type of atom or molecule.

The following diagrams show the molecules in two pure substances before mixing, and the mixture of molecules afterwards.

Look at the diagrams closely, and label each of them as either a single substance, or a mixture.

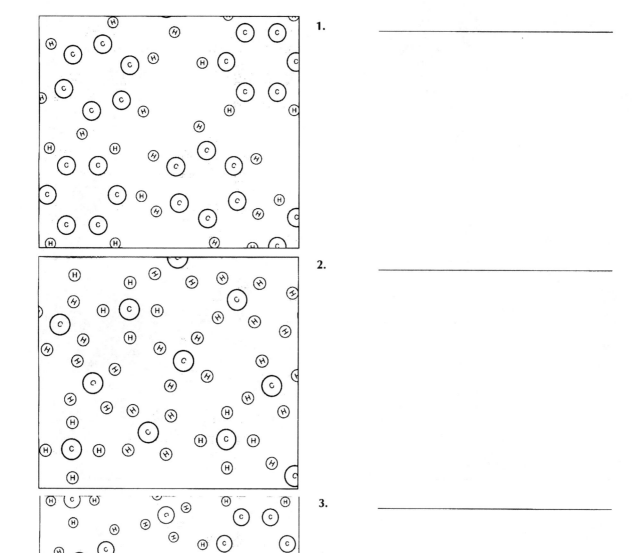

1. _____

2. _____

3. _____

RS•C

Here are some other diagrams showing mixtures:

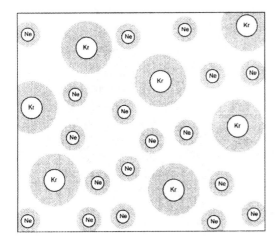
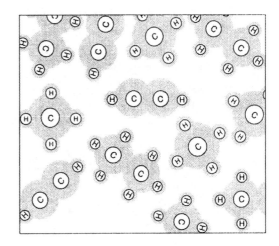

The following two diagrams show a single substance and a mixture. Complete the labels to show you know which is which:

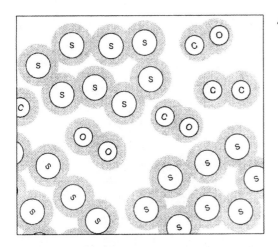

4. This diagram shows a

_____

because there is more than one type of

_____

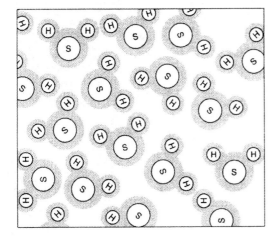

5 This diagram shows a

_____

because there is only one type of

_____

Two types of substance - element and compounds

To tell the difference between single substances and mixtures you need to be able to recognise diagrams of different types of atoms and molecules.

But scientists divide single substances into two types: elements and compounds. To spot the difference between elements and compounds you have to look more closely at the atoms or molecules.

Atoms are made of one core surrounded by a 'cloud' of electrons. Molecules are made of two or more cores surrounded by a 'cloud' of electrons.

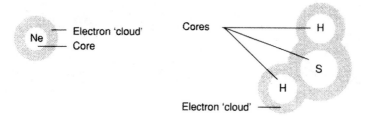

The atoms or molecules that form an element have only one type of core. (Scientists use different letters to represent the different elements).

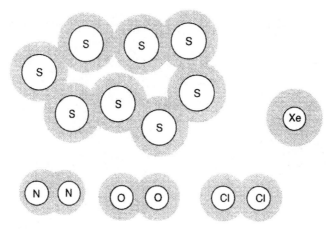

Atoms or molecules that have one type of core

Elements, compounds or mixtures? (2) page 6 of 8

RS•C

In a compound the molecules have two or more different types of core.

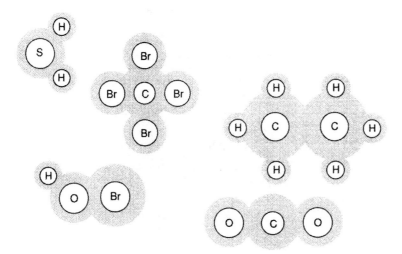

The following two diagrams show a molecule of an element and a molecule of a compound. Complete the labels to show you know which is which:

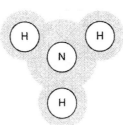

6. This diagram shows

_____

because there is more than one type of

_____

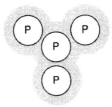

7 This diagram shows

_____

because there is only one type of

_____

To summarise:

Looking at:	If the same:	If different:
The types of single uncombined atoms	A single substance	A mixture
The types of molecules	A single substance	A mixture
The types of cores in a single molecule	A molecule of an element	A molecule of a compound

The following diagrams show molecules in a mixture, element and compound. Complete the labels to show you know which is which:

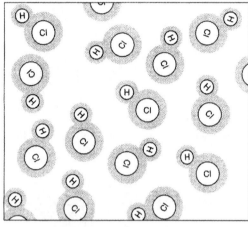

8. _____

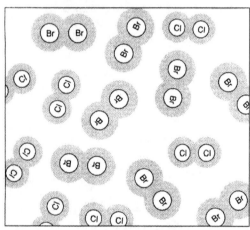

9. _____

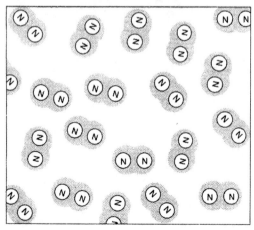

10. _____

RS•C

Elements, compounds or mixtures? (3)

Each of the following diagrams show the particles in a material. For each diagram, write whether you think it represents an element, a compound or a mixture

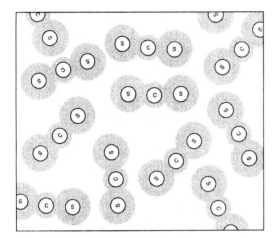

1. _____

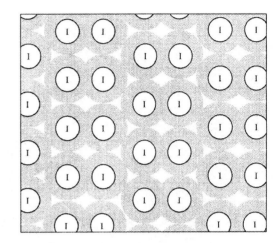

2. _____

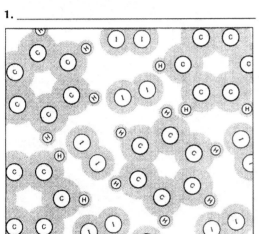

3. _____

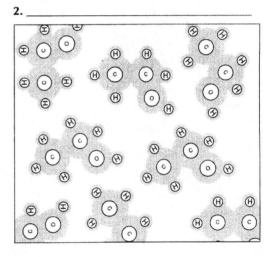

4. _____

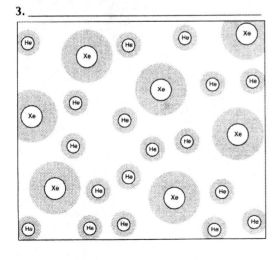

5. _____

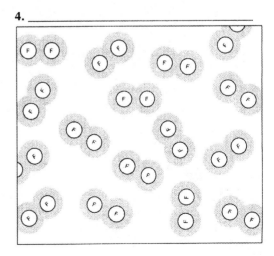

6. _____

RS•C

Elements, compounds or mixtures? (3) – page 1 of 1

RS•C

This page has been intentionally left blank.

RS•C

RS•C

Mass and dissolving

Target level

This exercise is primarily aimed at the 11–14 age range, although it may also be used to check the understanding of 14–16 year old students.

Topic

Rationale

Younger students may be satisfied with the idea that solutes 'disappear' when they dissolve, and even when the process of dissolving is appreciated students may not expect the mass of the solute to register in any measurements. This exercise asks students to predict the masses of solutions from given masses of solute and solvent, and also asks students to explain what happens to the solute, and to explain the emergent properties of the solution.

These ideas are discussed in Chapter 6 of the Teachers' notes.

When this exercise was piloted in schools it was found that some students did not expect mass to be conserved on dissolving (even though most recognised that the solute was still present in some form), and that students who conserved mass in their responses often had only vague ideas about how the properties of solutions arose.

The exercise was described by teachers as 'very good' and 'easy to follow', and was considered to be an effective diagnostic tool. It was suggested that the exercise could be used as an introductory activity before formally teaching about the topic at this level.

Instructions

It is suggested that this exercise may be used as a prelude to classroom discussion of the answers, although some teachers may wish to formally 'mark' students' responses. The precise level of an acceptable response will clearly depend upon the age and nature of the group.

Resources

■ Student worksheet
 – Mass and dissolving

Feedback

A suggested answer sheet is provided for teachers.

RS•C

Mass and dissolving – answers

1. Sugar and water

a) 210, 210

b) The sugar dissolved - it is still present, but as part of the solution.

(The molecules of sugar, which are much too small to be seen, are mixed with the molecules of water.)

2. Salt and water

a) 160, 160

b) The salt dissolved, and is now part of the solution. The salt particles are mixed with the water particles in the solution.

Note: If students have studied ionic bonding then it is worth emphasising that the sodium ions and chloride ions are separate in the mixture.

3. Copper sulfate and water

a) 255, 255

b) The blue colour is a property of the particles in the copper sulfate. The water turned blue as the copper sulfate dissolved to give a solution. The copper sulfate particles are mixed with the water particles. As the copper sulfate particles are spread throughout the solution the whole solution looks blue.

c) The copper sulfate dissolved - it is still present, but as part of the solution.

(The particles of copper sulfate, which are much too small to be seen, are mixed with the molecules of water.)

Note 1. Students commonly believe that the properties of a substance are due to its particles having that same property (see Chapter 6 of the Teachers' notes). This idea is usually not correct, so although in this case the colour may be seen to be a property of both the particles and the bulk material, the teacher should consider emphasising that this is unusual, and that most bulk properties are not shared by the particles.

Note 2. If students have studied ionic bonding then it is worth emphasising that the copper ions and sulfate ions are separate in the mixture. (The colour is due to the hydrated copper(II) ions.)

4. Particles in sugar and water

The liquid tastes sweet because the molecules of sugar are dissolved in the solution. Sugar has a sweet taste, so the solution tastes sweet because it contains the sugar molecules.

Mass and dissolving

This exercise is about what happens when solids dissolve in liquids.

1. Sugar and water

Some water was placed in a beaker, and its mass was measured using a balance. The mass of beaker and water was 200 g. Then 10 g of sugar was weighed out. The sugar was added to the water, and sank to the bottom. 10 minutes later the sugar could not be seen.

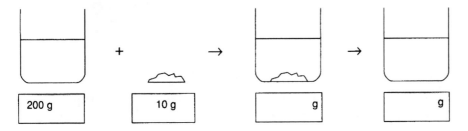

a) Fill in the boxes to show what you think the mass of the beaker and its contents would be when the sugar was first added, and then after it could no longer be seen.

b) Where did the sugar go? Explain your answer.

_____

_____

_____

_____

2. Salt and water

Some water was placed in a beaker, and its mass was measured using a balance. The mass of beaker and water was 150 g. Then 10 g of salt was weighed out. The salt was added to the water, and sank to the bottom. 10 minutes later the salt could not be seen.

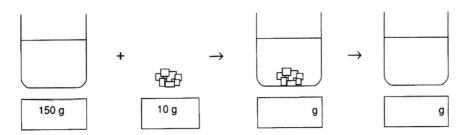

a) Fill in the boxes to show what you think the mass of the beaker and its contents would be when the salt was first added, and then after the salt could no longer be seen.

b) Where did the salt go?

_____

_____

_____

_____

RS•C

3. Copper sulfate and water

Some water was placed in a beaker, and its mass was measured using a balance. The mass of beaker and water was 250 g. Then 5 g of blue crystals of copper sulfate was weighed out. The copper sulfate was added to the water, and sank to the bottom. 20 minutes later the copper sulfate could not be seen, but the liquid had turned blue.

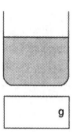

a) Fill in the boxes to show what you think the mass of the beaker and its contents would be when the copper sulfate was first added, and when it could no longer be seen.

b) Why did the water turn blue?

_____

_____

_____

_____

_____

c) Where did the copper sulfate go?

_____

_____

_____

4. Particles in sugar and water

The diagrams below represent the particles present at the different stages when sugar is dissolved in water. Not all the particles are shown.

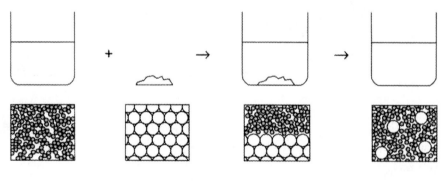

Why does the liquid taste sweet when sugar is added to water?

_____

_____

_____

_____

RS•C

RS•C

Changes in chemistry

Target level

This probe is intended for use with 11–14 year old students who have been taught the distinction between chemical and physical change. It may also be used with 14–16 year old students to check their knowledge.

Topics

Chemical and physical change.

Rationale

The distinction between chemical and physical change is not absolute, and there are examples of changes which teachers find difficult to classify (see Chapter 2 of the Teachers' notes). Students are often expected to distinguish between chemical and physical changes early in their study of chemistry, but some find this quite difficult. The probe provided here asks students to explain what they mean by chemical and physical change, to classify three examples and explain their reasons. Particle diagrams for 'before' and 'afterwards' are provided to help the students.

These ideas are discussed in Chapter 6 of the Teachers' notes.

During piloting, teachers suggested that this 'could be a useful teaching tool' that helped 'to clarify the idea in some pupils' minds'. It was felt that the use of particle diagrams was helpful. Students were reported to enjoy the activity, and found it easy to understand what to do.

One teacher commented that the activity 'helped [students] to grasp the differences but they found it hard to define the changes'. It might be appropriate to suggest to students who are not sure what to put for the definitions (at the start of the probe) to move on to the examples, and return to the definitions later.

Resources

■ Student worksheet
  – Changes in chemistry

Feedback for students

A suggested answer sheet for teachers is provided. When providing feedback on this probe teachers should bear in mind the difficulties of defining the physical/chemical change distinction, as discussed in Chapter 2 of the Teachers' notes.

RS•C

Changes in chemistry – answers

Alternative reasons that may be accepted are given in parentheses

1. A physical change is a change where no new substance is produced.

(....a change that does not involve the breaking/forming of strong chemical bonds.)

(.....a change where molecules/ions etc are rearranged, but not changed.)

2. A chemical change is a change where a new substance is produced.

(.....a change that involves the breaking/forming of strong chemical bonds.)

(.....a change where new molecules/ions etc are formed.)

3. Physical change:

No new substance is produced

(the same molecules are present before and after the change)

(the change may readily be reversed)

(the energy change involved is modest)

4. Chemical change:

A new substance is formed

(strong chemical bonds are broken – *eg* in the oxygen molecules - and new chemical bonds are formed in the metal oxide)

(different particles are present after the change - oxide ions rather than oxygen molecules)

(this change is not easily reversed)

(a great deal of energy is often given out in this change)

5. Physical change:

No new substance is formed (NB the solution is not a pure substance, but a mixture)

(the same particles are present after the change as before)

(this reaction is readily reversed – by evaporation)

(the energy change for dissolving is minimal)

Note, however, that the ionic bonds in the lattice have been disrupted, which may suggest dissolving could be considered as a chemical change.

Note: Energy changes are not the best way to characterize these changes.

Changes in chemistry

In science we describe the changes that occur to substances as either physical changes or chemical changes. Explain what you think these terms mean:

1. A physical change is

_____

_____

_____

2. A chemical change is

_____

_____

_____

Below and over the page you will find three examples of substances being changed. The diagrams show some of the molecules or other particles before and after the change. For each example:

■ decide whether the change is physical or chemical, and

■ try to explain your reasons.

3. Some very cold liquid nitrogen is cooled even further, until it freezes:

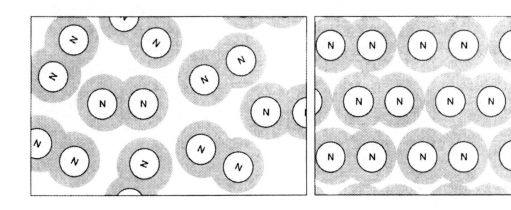

before   after

This is a _____ change because

_____

_____

_____

4. Some magnesium is heated in oxygen until it burns:

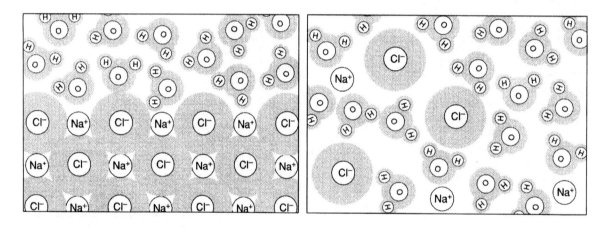

before   after

This is a _____ change because

_____

_____

_____

5. Some sodium chloride is added to a beaker of water, and left to dissolve:

before   after

This is a _____ change because

_____

_____

_____

RS•C

RS•C

Revising acids

Target level

This exercise is designed for 11–14 year old students who have studied the topic of acids.

Topic

Acids and alkalis.

Rationale

Concept mapping is a useful technique which can help the teacher identify students' alternative conceptions, and evaluate the extent and degree of integration of students' knowledge of a topic. It also provides an alternative to linear notes and summaries that will appeal to some students (either 'for a change' of activity, or because of their preferred learning styles), and which encourages active processing of information (see Chapter 5 of the Teachers' notes), and improves study skills by getting students to think about their own learning. The uses and features of concept mapping are discussed in Chapter 3 of the Teachers' notes. This exercise is designed to be accessible to students (and teachers) who are not already experienced at concept mapping. The exercise is provided at three levels of difficulty (Foundation, Standard and Extension).

The Foundation level version was designed to provide an achievable task for students will little or no knowledge of the topic – provided that they are able to deal with the language demands. When the materials were piloted some teachers and students suggested that this was too easy: however, many students tackling the Standard level task made significant errors, and it is suggested that the Foundation task may be a useful learning activity even though it will provide little assessment information.

The materials were judged to be 'very useful' both as a revision activity, and as an introduction to a study technique. The materials led to useful classroom discussion. As might be expected, student views varied; some preferred this approach and found it useful, where others found it difficult and preferred more familiar revision techniques.

Instructions

The materials required by students depend upon the level of demand required.

Foundation level: students will require

- **Acid revision map** – worksheet with concept map, separate worksheets will be required for each exercise.

- **Labelling the revision map** – worksheet providing labels to be matched to map links. This could be adapted to provide fewer labels.

Standard level: students will require

- **Acid revision map** – worksheet with concept map.

- **Completing the revision map labels** – worksheet providing incomplete map labels. This could be adapted to provide fewer labels. Teachers may wish to check students' responses before they commence labelling the map.

RS•C

Extension level: students will require

■ **Outline acid revision map** – worksheet with outline concept map.

■ **Connecting up the revision map** - worksheet giving instructions for competing the map.

Teachers should decide whether or not to explain the meaning of the arrows in the concept map.

Resources

■ Student worksheets
  – Acid revision map
  – Labelling the revision map
  – Completing the revision map labels
  – Outline acid revision map
  – Connecting up the revision map
  – Example concept map - acids

Feedback for students

An example of a completed map is provided, **Example concept map – acids**, and teachers may wish to issue this to students once their own versions have been completed and discussed. It should be pointed out that there is no one correct map, and teachers may wish to get students to add any relevant additional features from students' maps to the example map. For instance, there is no connection between alkali and base on the map.

RS•C

Acid revision map

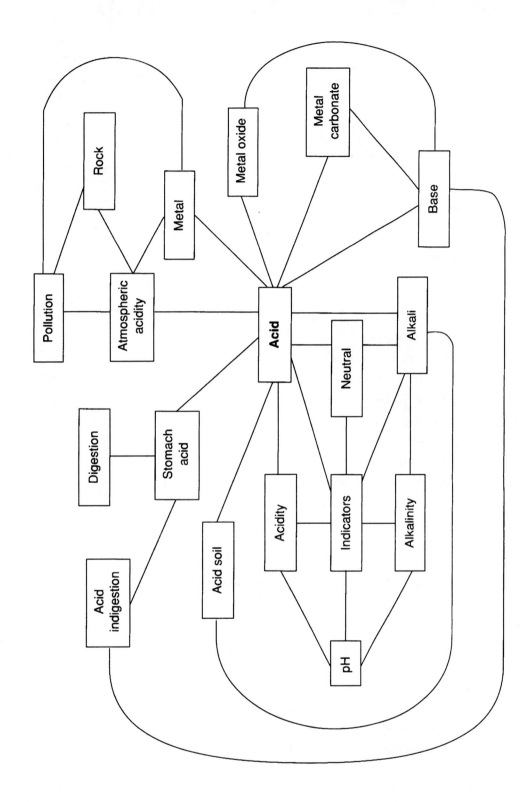

Labelling the revision map

You have been given a copy of the acid revision map. This shows some of the important ideas you may have met when you studied acids and bases in your science class. Each line on the map stands for an idea that could be put into a sentence.

The links are not explained on the map. Read through the statements below, and work out which link on the map each sentence is about.

Label each line on the map with the number of the statement – eg

1. Acidity is a property of acids.
2. Acids can be identified using indicators.
3. Acidity can be measured using the pH scale.
4. Acidity can be detected using an indicator.
5. Alkalinity is a property of alkalis.
6. Alkalinity can be detected using an indicator.
7. Alkalinity can be measured using the pH scale.
8. Neutral solutions can be identified using indicators.
9. Alkalis can be identified using indicators.
10. Acids are not neutral solutions.
11. Alkalis are not neutral solutions.
12. pH may be found using universal indicator.
13. Acids react with alkalis to give a salt and water.
14. Bases react with acids.
15. An alkali is a base which dissolves in water.
16. Metal carbonates are bases.
17. Metal carbonates react with acids to give a salt and carbon dioxide.
18. Metal oxides are bases.
19. Metal oxides react with acids to give salts and water.
20. Some metals react with acid to give a salt and hydrogen.
21. Acids in the air cause atmospheric acidity.
22. Atmospheric acidity is increased by some forms of pollution.
23. Atmospheric acidity causes weathering of rocks.
24. Pollution can increase the rate of weathering of rock.
25. Atmospheric acidity causes the corrosion of some metals.
26. Pollution can increase the rate of corrosion of metals.
27. Acid is found in the stomach.
28. Stomach acid helps us digest our food.
29. Too much stomach acid can cause indigestion.
30. Some bases are used to relieve acid indigestion.
31. Some soils contain too much acid for most plants to grow.
32. An alkali is sometimes added to soil to neutralise acidity.

RS•C

Completing the revision map labels

You have been given a copy of the acid revision map. This shows some of the important ideas you may have met when you studied acids and bases in your science class. Each line on the map stands for an idea that could be put into a sentence.

The links are not explained on the map. Read through the statements below, and work out which link on the map each sentence is about.

However, each sentence has a key word or phrase missing – so you will also need to complete the sentences!

Label each line on the map with the letter of the statement – *eg*

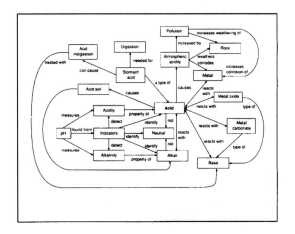

A. Acids in the _____ cause atmospheric acidity.

B. Atmospheric acidity is increased by some forms of _____.

C. _____ _____ causes weathering of rocks.

D. _____ can increase the rate of weathering of rock.

E. Atmospheric acidity causes the corrosion of some _____.

F. _____ can increase the rate of corrosion of metals.

G. _____ is found in the stomach.

H. _____ _____ helps us digest our food.

I. Too much stomach acid can cause _____.

J. Some bases are used to relieve _____ _____.

K. Some _____ contain too much acid for many plants to grow.

L. _____ is sometimes added to soil to neutralise acidity.

M. Acids react with _____ to give a salt and water.

N. Bases react with _____.

O. An _____ is a base which dissolves in water.

P. Metal carbonates are _____.

Q. ____ _____ react with acids to give a salt and carbon dioxide.

R. Metal oxides are _____.

S. Metal oxides react with _____ to give salts and water.

T. Some _____ react with acid to give a salt and hydrogen.

U. Acidity is a property of _____.

V. Acids can be identified using _____.

W. Acidity can be measured using the _____ scale.

X. Acidity can be detected using an _____.

Y. Alkalinity is a property of _____.

Z. Alkalinity can be detected using an _____.

α Alkalinity can be measured using the _____ scale.

Ω Neutral solutions can be identified using _____.

δ Alkalis can be identified using _____.

Φ Acids are not _____ solutions.

Σ Alkalis are not _____ solutions.

ψ pH may be found using universal _____.

RS•C

Outline acid revision map

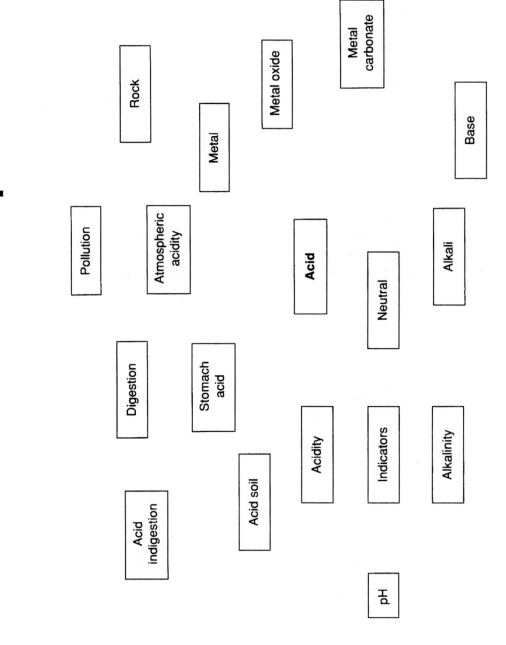

Connecting up the revision map

You have been given a copy of an outline of a revision map for the topic of acids. This shows some of the things you may have met when you studied acids and bases in your science class. However the map is not complete!

The boxes on the map need to be connected to show how the ideas are linked.

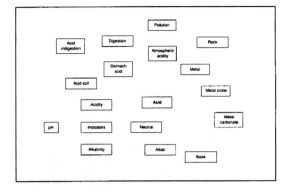

Instructions

1. Look at the outline map. Find two boxes that you think you can connect.

2. Draw a clear line between the two boxes.

3. Add a label to the line to explain the connection.

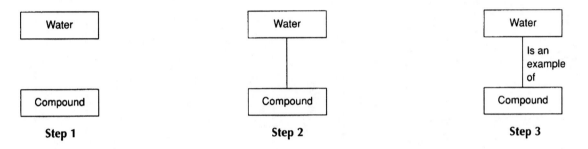

| Step 1 | Step 2 | Step 3 |

4. Repeat for as many connections as you can find.

5. See if you can think of any other boxes that would fit on this revision map. Draw them in.

6. Show the connections for the new boxes in the same ways as above (steps 2 and 3).

Example concept map – acids

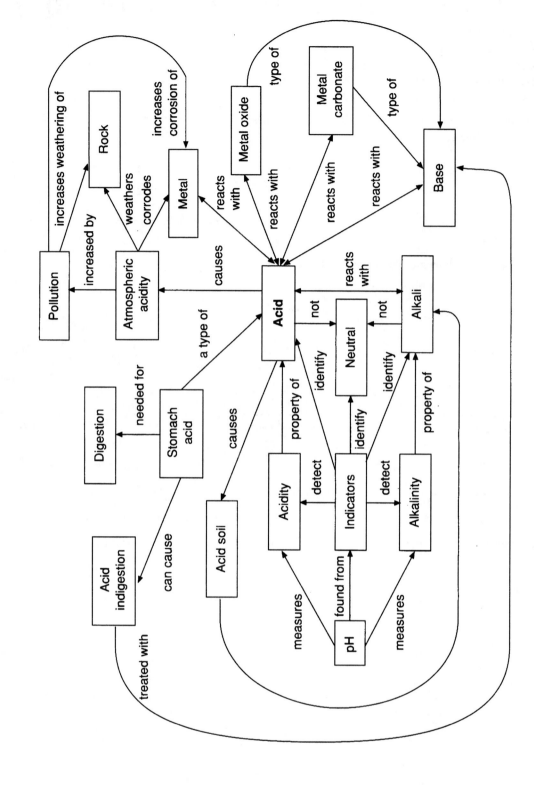

RS•C

RS•C

Word equations

Target level

These materials are designed for use with 11–14 year old students who have been taught to use word equations. The materials will also be useful for 14–16 year olds students who need to revisit this topic.

Topic

Using word equations to represent chemical reactions.

Rationale

Many students find it difficult to write word equations, which require an appreciation of the nature of chemical change (including conservation of matter), and familiarity with chemical names and the patterns of common reaction types. These materials provide probes for exploring whether students can complete word equations, and a set of practice exercises.

These ideas are discussed in Chapter 9 of the Teachers' notes.

In the pilot teachers judged the materials 'excellent', 'very useful' and 'helpful for revision'. Some teachers found the responses of some of their pupils to be 'disappointing' (or even 'shocking'). The probes were thought to provide an interesting 'look into [students'] minds' and to lead to useful classroom discussion.

Although teachers found it useful that students were asked to give reasons for their answers, some of the students did not like having to try to explain their reasons. (Some teachers may wish to ask students to just complete the equations in the probes, and to leave the spaces for making notes when going through the answers.) Students were reported to find the materials helpful and easy to follow, and were considered to have greater understanding afterwards.

Instructions

These materials may either be used with students who should have mastered word equations as a pre-test (to identify students needing practice), a remedial exercise, and a post-test; or as end-of-topic review material with students meeting word equations for the first time.

The materials include:

■ **Completing word equations (1)**
A probe to explore whether students can complete word equations.

■ **Word equations**
A study exercise including an information sheet providing practice in completing word equations for five common types of reaction. This exercise could be set as a private study (homework) task.

■ **Completing word equations (2)**
A probe to explore whether students can complete word equations.

RS•C

Resources

■ Student worksheets
  – Completing word equations (1)
  – Word equations   – Information sheet
                      – The reaction between acids and alkalis
                      – The reaction between acids and metals
                      – The reaction between acids and metal carbonates
                      – Displacement reactions
                      – Synthesis reactions
  – Completing word equations (2)

Feedback

Teachers' answer sheets are included for each of the probes, and the study activity.

Word equations – answers

Completing word equations (1)

Teachers will have their own ideas about what is an acceptable explanation for students' completions, depending upon the age of the students and the depth to which ideas about reactions have been studied. Suggested explanations are provided.

1. nitric acid + potassium hydroxide → **potassium nitrate** + water
   In this type of reaction a salt is formed from an acid and an alkali.
   The reactants include elements which must also be present in the products.
   As the reacting alkali is potassium hydroxide, then the product will be a potassium salt.
   As the reacting acid is nitric acid, then the salt produced will be a nitrate.

2. zinc + **copper nitrate** → zinc nitrate + copper
   In this type of reaction a more reactive metal displaces a less reactive metal from its salt.
   The reactants include elements which must also be present in the products.
   As the displaced metal is copper, then the reacting salt must have been a copper salt.
   As the product was a nitrate, then the reacting salt will also be a nitrate.

3. **sulfuric acid** + zinc carbonate → zinc sulfate + water + carbon dioxide
   In this type of reaction an acid reacts with a carbonate to give a salt, water and carbon dioxide.
   The reactants include elements which must also be present in the products.
   As the product is a sulfate then the reacting acid must be sulfuric acid.

4. calcium + chlorine → **calcium chloride**
   In this type of reaction a binary compound is prepared from two elements.
   The reactants include elements which must also be present in the products.

5. magnesium + hydrochloric acid → **magnesium chloride**
   In this type of reaction an acid reacts with a metal to produce a salt and hydrogen.
   The reactants include elements which must also be present in the products.
   As the reacting acid is hydrochloric acid, then the salt produced will be a chloride.
   As the reacting metal is magnesium, then the product must be a magnesium salt.

Word equations – the reaction between acids and alkalis

1. Potassium chloride
   Sodium nitrate

2. Potassium hydroxide

Word equations – the reaction between acids and metals

1. Iron nitrate
   Zinc sulfate
   Magnesium chloride, Magnesium nitrate

2. Calcium

RS•C

Word equations – the reaction between acids and metal carbonates

1. Copper chloride, Copper sulfate
   Zinc chloride, Zinc nitrate
   Iron nitrate, Iron sulfate

2. Magnesium chloride

Word equations – Displacement reactions

1. Iron

2. Zinc sulfate

3. Magnesium, copper

4. Copper sulfate

5. Zinc chloride, iron

6. Iron sulfate

7. Zinc, copper nitrate

8. Copper

9. Iron nitrate, copper

It may be worth pointing out that the order of the reactants or products in the equation is irrelevant.

Word equations – synthesis reactions

1. Sulfur
   Fluoride

2. Magnesium sulfide, Magnesium chloride
   Iron fluoride, Iron chloride
   Zinc oxide, Zinc sulfide, Zinc fluoride
   Copper oxide, Copper sulfide, Copper chloride

3. Sodium iodide
   Nickel sulfide
   Calcium

Completing word equations (2)

Teachers will have their own ideas about what is an acceptable explanation for students' completions, depending upon the age of the students and the depth to which ideas about reactions have been studied. Suggested explanations are provided.

1. sulfuric acid + sodium hydroxide → **sodium sulfate** + water
   In this type of reaction a salt is formed from an acid and an alkali.
   The reactants include elements which must also be present in the products.
   As the reacting alkali is sodium hydroxide, then the product will be a sodium salt.
   As the reacting acid is sulfuric acid, then the salt produced will be a sulfate.

2. iron + **copper chloride** → iron chloride + copper
   In this type of reaction a more reactive metal displaces a less reactive metal from its salt.
   The reactants include elements which must also be present in the products.
   As the displaced metal is copper, then the reacting salt must have been a copper salt.
   As the product was a chloride, then the reacting salt will also be a chloride.

RS•C

3. **hydrochloric** acid + magnesium carbonate → magnesium chloride + water + carbon dioxide
   In this type of reaction an acid reacts with a carbonate to give a salt, water and carbon dioxide.
   The reactants include elements which must also be present in the products.
   As the product is a chloride, then the reacting acid must be hydrochloric acid.

4. zinc + oxygen → **zinc oxide**
   In this type of reaction a binary compound is prepared from two elements.
   The reactants include elements which must also be present in the products.

5. calcium + nitric acid → **calcium nitrate** + hydrogen
   In this type of reaction an acid reacts with a metal to produce a salt and hydrogen.
   The reactants include elements which must also be present in the products.
   As the reacting acid is nitric acid, then the salt produced will be a nitrate.
   As the reacting metal is calcium, then the product must be a calcium salt.

Completing word equations (1)

Word equations are used to describe chemical reactions. Look at the word equations below. In each case complete the word equation by adding the name of the missing substance. (Explain your answers.)

1. nitric acid + potassium hydroxide → _____ + water

    I think this is the answer because

    _____

    _____

    _____

    _____

2. zinc + _____ → zinc nitrate + copper

    I think this is the answer because

    _____

    _____

    _____

    _____

3. _____ + zinc carbonate → zinc sulfate + water + carbon dioxide

    I think this is the answer because

    _____

    _____

    _____

    _____

4. calcium + chlorine → _____

    I think this is the answer because

    _____

    _____

    _____

5. magnesium + hydrochloric acid → _____ + hydrogen

    I think this is the answer because

    _____

    _____

    _____

    _____

Word equations – information sheet

Word equations are a shorthand used to describe chemical reactions.

Although there are many millions of possible chemical reactions you are not expected to know about them all.

It is useful to remember that many reactions are of similar types.

1. Metal + acid

One type of reaction is that between a metal and an acid. When a metal reacts with an acid the reaction produces a salt, and hydrogen gas is released. The salt produced depends upon the metal and the acid. If magnesium reacts with hydrochloric acid, then the salt produced is magnesium chloride.

metal + acid → salt + hydrogen

2. Metal carbonate + acid

Metal carbonates also react with acid, to give a salt. When a carbonate reacts with acid the gas carbon dioxide is given off. The salt produced depends upon which acid, and which metal carbonate react. If zinc carbonate reacts with sulfuric acid, then the salt produced is zinc sulfate.

metal carbonate + acid → salt + carbon dioxide + water

3. Acid + alkali

When an alkali and an acid react the product is a salt solution. The particular salt produced depends upon which acid and which alkali reacted. When nitric acid reacts with potassium hydroxide the salt produced is potassium nitrate.

acid + alkali → salt + water

4. Metal + salt solution

When a reactive metal is placed in the solution of a salt of a less reactive metal, a 'displacement' reaction occurs. The more reactive metal is said to displace the less reactive metal from solution. For example zinc is added to copper nitrate solution the copper is displaced and the solution will contain zinc nitrate.

5. Element + element

When two elements react together to form a compound the compound is given a name to show which elements reacted. So if calcium reacts with chlorine the compound is called calcium chloride.

These examples show you the patterns that are found in five important types of reaction. If you look for patterns you should find it easier to work out how to complete word equations.

In this worksheet you will find some examples to help you practice thinking about word equations.

Word equations – the reaction between acids and alkalis

When an acid reacts with an alkali, a salt and water is produced:

acid + alkali → salt + water

for example

hydrochloric acid + sodium hydroxide → sodium chloride + water

The salt that is produced depends upon which acid and which alkali react. The following table provides a summary of the name of the salt produced by different reactions between acids and alkalis.

1. Complete the table (Hint – look for the patterns)

Name of acid	Name of alkali	
	Sodium hydroxide	Potassium hydroxide
Hydrochloric acid	Sodium chloride	
Nitric acid		Potassium nitrate
Sulfuric acid	Sodium sulfate	Potassium sulfate

Salts produced when acids react with alkalis

2. Complete the following word equation

(acid + alkali → salt + water)

nitric acid + _____ → potassium nitrate + water

Word equations – the reaction between acids and metals

When an acid reacts with metal, a salt and hydrogen are produced:

acid + metal→ salt + hydrogen

for example

nitric acid + calcium → calcium nitrate + hydrogen

The salt that is produced depends upon which acid and which metal react. The following table provides a summary of the name of the salt produced by different reactions between acids and metals.

1. Complete the table (Hint – look for the patterns)

Name of metal	Name of acid		
	Hydrochloric acid	**Nitric acid**	**Sulfuric acid**
Iron	Iron chloride		Iron sulfate
Zinc	Zinc chloride	Zinc nitrate	
Magnesium			Magnesium sulfate

2. Complete the following word equation

(acid + metal → salt + hydrogen)

hydrochloric acid + _____ → calcium chloride + hydrogen

Word equations – the reaction between acids and metal carbonates

When an acid reacts with a metal carbonate, a salt, water and carbon dioxide are produced:

Acid + metal carbonate → salt + water + carbon dioxide

for example

sulfuric acid + zinc carbonate → zinc sulfate + water + carbon dioxide

The salt that is produced depends upon which acid and which metal react. The following table provides a summary of the name of the salt produced by different reactions between acids and metal carbonates.

1. Complete the table (Hint – look for the patterns)

Name of metal carbonate	Name of acid		
	Hydrochloric acid	Nitric acid	Sulfuric acid
Copper carbonate		Copper nitrate	
Zinc carbonate			Zinc sulfate
Iron carbonate	Iron chloride		

2. Complete the following word equation

(acid + metal carbonate → salt + water + carbon dioxide)

hydrochloric acid + magnesium carbonate → _____ + water + carbon dioxide

Word equations – displacement reactions

When a reactive metal is added to a solution containing the salt of a less reactive metal, a reaction occurs.

See if you can complete the following word equations (Hint – look for the patterns).

eg magnesium + iron chloride → magnesium chloride + iron

1.   magnesium + iron nitrate → magnesium nitrate + _____

2.   magnesium + _____ → magnesium sulfate + zinc

3.   _____ + copper sulfate → magnesium sulfate + _____

4.   zinc + _____ → zinc sulfate + copper

5.   zinc + iron chloride → _____ + _____

6.   zinc + _____ → zinc sulfate + iron

7.   _____ + _____ → zinc nitrate + copper

8.   iron + copper sulfate → iron sulfate + _____

9.   iron + copper nitrate → _____ + _____

Word equations – synthesis reactions

When a metallic element reacts with a non-metallic element a compound is produced.

The name of the compound contains the name of the metal and the altered name of the non-metal.

1. Complete this summary:

When oxygen reacts with a metal, the compound is called an oxide.

When chlorine reacts with a metal, the compounds is called a chloride.

When _____ reacts with a metal, the compound is called a sulfide.

When fluorine reacts with a metal, the compound is called a _____.

2. Complete the table below: (Hint – look for the patterns)

Name of metal	Name of non-metal			
	Oxygen	Sulfur	Fluorine	Chlorine
Magnesium	Magnesium oxide		Magnesium fluoride	
Iron	Iron oxide	Iron sulfide		
Zinc				Zinc chloride
Copper			Copper fluoride	

3. Complete the following word equations:

sodium + iodine → _____

nickel + sulfur → _____ _____

_____ + bromine → calcium bromide

RS•C

Completing word equations (2)

Word equations are used to describe chemical reactions. Look at the word equations below. In each case complete the word equation by adding the name of the missing substance. (Explain your answers.)

1. sulfuric acid + sodium hydroxide → _____ + water

   I think this is the answer because

   _____

   _____

   _____

   _____

2. iron + _____ → iron chloride + copper

   I think this is the answer because

   _____

   _____

   _____

   _____

3. _____ acid + magnesium carbonate → magnesium chloride + water + carbon dioxide

   I think this is the answer because

   _____

   _____

   _____

   _____

4. zinc + oxygen → _____

   I think this is the answer because

   _____

   _____

   _____

   _____

5. calcium + nitric acid → _____ + hydrogen

   I think this is the answer because

   _____

   _____

   _____

   _____

RS•C

This page has been intentionally left blank.

RS•C

RS•C

Definitions in chemistry

Target level

This probe is primarily designed for discussion work with students taking post-16 chemistry courses, or for more able students in the 14–16 age range.

Topics

Definitions of basic terms in chemistry: element, compound, atom, molecule.

Rationale

Even the most basic and common chemical terms may not have clear and unambiguous definitions (see Chapter 2 of the Teachers' notes). Students are often found to be confused over the most basic chemical concepts, even after some years of familiarity through science lessons. (These ideas are discussed in Chapter 6 of the Teachers' notes.) This exercise gives students an opportunity to discuss and give a critique of a selection of definitions (from various sources) of the most basic chemical terms, thus exploring and revealing their own understanding of these concepts.

The exercise does require a fairly high level of linguistic competence, and would not be suitable for all students in the 14–16 age range if set as an independent activity.

During piloting, it was found that most students found it helpful and appreciated the approach. It was also found to be a very useful exercise for the teacher (in terms of revealing students' ideas), and considered to be an admirable approach for teaching definitions.

Instructions

If the activity is set as group work the teacher may wish to issue only one worksheet per group to ensure that there is discussion towards a consensus view.

Resources

■ Student worksheet
   – Definitions in chemistry

Feedback for students

A series of discussion points are provided which highlight the strengths and weaknesses of the definitions provided.

RS•C

Definitions in chemistry – answers

Discussion points

The aim in this exercise is partly to spot incorrect definitions, but also to highlight the limitations of the standard textbook statements, which may not be that helpful to students. Which box is ticked is less important than the students' reasoning for their choice.

1. Element:

   a) A substance that is made of only one kind of atom. The meaning of 'kind' of atom must here apply to atomic/proton number. (The definition does not allow for existence of isotopes.)

   b) A substance which cannot be split up into simpler substances. This definition requires one to be able to recognise a 'simpler' substance.

2. Compound:

   a) Is made of two elements mixed together. A compound can be more than two elements – and they are not mixed.

   b) A substance consisting of atoms of different elements joined together. Does NaCl (for example) consist of atoms joined?

   c) A chemical substance made up of two or more elements bonded together, so that they cannot be separated by physical means. This definition relies upon a clear understanding of what are 'physical' means.

   d) A product which has properties different from those of either of the component substances and which is formed with an accompanying energy change is called a compound. This implies that the compound is comprised of only two elements. The definitions also rely upon the word 'product' being used here in a technical sense. There are many mixtures of substances (which in everyday terms might be call 'products' of mixing) which seem to have different properties from the component substances.

3. Atom:

   a) The simplest structure in chemistry. It contains a nucleus with protons and neutrons, and electrons moving around in shells. The nucleus is a simpler structure than the atom. (Some students may know that individual nucleons also have structure - and are therefore simpler structures still.) The sodium cation is a simpler structure than the sodium atom. Of course, the atom is the simplest structure in chemistry, if we decide that any simpler structures belong to physics!

   b) The smallest part of an element which can exist as a stable entity. Stability can be in terms of interacting with other chemical species, or in terms of spontaneous decay. For most elements separate atoms are labile, and will soon interact with other chemical species. For non-metals the molecule is the smallest stable part in a chemical sense. In terms of spontaneous decay, most atoms are stable, but so are most nuclei met in chemistry; and so are individual electrons, and individual protons (although not individual neutrons) - which are clearly smaller parts.

c) The building blocks of life, the Lego® of nature. An interesting metaphor - although most chemical structures are not formed of discrete atoms, but of various arrangements of atomic cores and electrons (see Chapter 7 of the Teachers' notes).

d) The smallest particle of an element that still shows the chemical properties of the element. Most chemical properties are not exhibited by individual atoms. An atom of sulfur does not share the chemical properties of sulfur. The smallest unit which could be claimed to have these properties is the molecule.

e) The smallest portion of an element that can take part in a chemical reaction. Although atoms may be take part in reactions (when there is a free radical mechanism), reactants are seldom present in atomic form, and the smallest portion of a non-metallic element should be considered to be the molecule.

f) Smallest particle that can be found. It is made up of protons, neutrons and electrons. Clearly particles smaller than atoms can be found with advanced scientific techniques. (In everyday terms, the smallest 'particle' that could be found by a student would be many orders of magnitude larger than the atom.)

g) The smallest particles that can be obtained by chemical means. This would seem to be a statement with little meaning. Few chemical processes produce atoms – products are normally in the form of ions or molecules etc. However the 'chemical means' of an electrolytic cell will provide a source of electrons - smaller particles than atoms.

4. Molecule:

a) The smallest particle of matter which can exist in a free state. It is unlikely that most students will appreciate what is meant by a 'free state', but most will have television sets which work due to a beam of electrons (smaller particles existing free of atoms *etc*).

b) Something that is formed by two atoms bonding together. Molecules often comprise of more than two atomic centres.

c) The smallest portion of a substance capable of existing independently and retaining the properties of the original substance. A molecule does not share all of the properties of a substance, as some derive from the larger scale arrangement of molecules.

d) Group of two or more atoms bonded together. A molecule of an element consists of one or more like atoms; a molecule of a compound consists of two or more different atoms bonded together. The second part of the sentence contradicts the first!

Definitions in chemistry

Definitions tell us what words mean. Good definitions can be very useful, but sometimes definitions can be wrong, or just confusing. To be helpful a definition needs to be correct, and to make sense.

Below are some definitions of important words in science. Some of the definitions below come from books, and some have been provided by young people studying science. Read each definition carefully and decide (a) if you think it is correct, and (b) whether it is a definition that would help someone learning about science. Try to explain your reasons, if you can.

1. Element: the following definitions have been given to the word 'element'

a)   **Element: A substance that is made of only one kind of atom.**

✔ Is the definition correct?   ✔ Would the definition help someone to learn?

❑ Yes, it is correct   ❑ Yes, it is helpful

❑ No, it is wrong   ❑ No, it is not helpful

❑ I am not sure   ❑ I am not sure

I think this because:

_____

_____

_____

_____

_____

b)   **Element: A substance which cannot be split up into simpler substances.**

✔ Is the definition correct?   ✔ Would the definition help someone to learn?

❑ Yes, it is correct   ❑ Yes, it is helpful

❑ No, it is wrong   ❑ No, it is not helpful

❑ I am not sure   ❑ I am not sure

I think this because:

_____

_____

_____

_____

_____

RS•C

2. Compound: the following definitions have all been given to the word 'compound'

a) **Compound: Is made of two elements mixed together.**

✔ Is the definition correct?

❑ Yes, it is correct

❑ No, it is wrong

❑ I am not sure

✔ Would the definition help someone to learn?

❑ Yes, it is helpful

❑ No, it is not helpful

❑ I am not sure

I think this because:

_____

_____

_____

_____

_____

b) **Compound: A substance consisting of atoms of different elements joined together.**

✔ Is the definition correct?

❑ Yes, it is correct

❑ No, it is wrong

❑ I am not sure

✔ Would the definition help someone to learn?

❑ Yes, it is helpful

❑ No, it is not helpful

❑ I am not sure

I think this because:

_____

_____

_____

_____

_____

c) **Compound: A chemical substance made up of two or more elements bonded together, so that they cannot be separated by physical means.**

✔ Is the definition correct?

❑ Yes, it is correct

❑ No, it is wrong

❑ I am not sure

✔ Would the definition help someone to learn?

❑ Yes, it is helpful

❑ No, it is not helpful

❑ I am not sure

I think this because:

_____

_____

_____

_____

_____

d) **Compound: A product which has properties different from those of either of the component substances and which is formed with an accompanying energy change is called a compound.**

✔ Is the definition correct?

❑ Yes, it is correct

❑ No, it is wrong

❑ I am not sure

✔ Would the definition help someone to learn?

❑ Yes, it is helpful

❑ No, it is not helpful

❑ I am not sure

I think this because:

_____

_____

_____

_____

_____

3. Atom: the following definitions have been given to the word 'atom'

a) **Atom: The simplest structure in chemistry. It contains a nucleus with protons and neutrons, and electrons moving around in shells.**

✔ Is the definition correct?

❑ Yes, it is correct

❑ No, it is wrong

❑ I am not sure

✔ Would the definition help someone to learn?

❑ Yes, it is helpful

❑ No, it is not helpful

❑ I am not sure

I think this because:

_____

_____

_____

_____

_____

b) **Atom: The smallest part of an element which can exist as a stable entity.**

✔ Is the definition correct?

❑ Yes, it is correct

❑ No, it is wrong

❑ I am not sure

✔ Would the definition help someone to learn?

❑ Yes, it is helpful

❑ No, it is not helpful

❑ I am not sure

I think this because:

_____

_____

_____

_____

_____

RS•C

c)   **Atom: The building blocks of life, the Lego® of nature.**

✔   Is the definition correct?   ✔   Would the definition help someone to learn?

❑   Yes, it is correct   ❑   Yes, it is helpful

❑   No, it is wrong   ❑   No, it is not helpful

❑   I am not sure   ❑   I am not sure

I think this because:

_____

_____

_____

_____

_____

d)   **Atom: The smallest particle of an element that still shows the chemical properties of the element.**

✔   Is the definition correct?   ✔   Would the definition help someone to learn?

❑   Yes, it is correct   ❑   Yes, it is helpful

❑   No, it is wrong   ❑   No, it is not helpful

❑   I am not sure   ❑   I am not sure

I think this because:

_____

_____

_____

_____

_____

e)   **Atom: The smallest particle of an element that can take part in a chemical reaction.**

✔   Is the definition correct?   ✔   Would the definition help someone to learn?

❑   Yes, it is correct   ❑   Yes, it is helpful

❑   No, it is wrong   ❑   No, it is not helpful

❑   I am not sure   ❑   I am not sure

I think this because:

_____

_____

_____

_____

_____

f) **Atom: Smallest particle that can be found. It is made up of protons, neutrons and electrons.**

✔ Is the definition correct? ✔ Would the definition help someone to learn?

❑ Yes, it is correct ❑ Yes, it is helpful

❑ No, it is wrong ❑ No, it is not helpful

❑ I am not sure ❑ I am not sure

I think this because:

_____

_____

_____

_____

_____

g) **Atom: The smallest particles that can be obtained by chemical means.**

✔ Is the definition correct? ✔ Would the definition help someone to learn?

❑ Yes, it is correct ❑ Yes, it is helpful

❑ No, it is wrong ❑ No, it is not helpful

❑ I am not sure ❑ I am not sure

I think this because:

_____

_____

_____

_____

_____

4. Molecule: the following definitions have been given to the word 'molecule'

a) **Molecule: The smallest particle of matter which can exist in a free state.**

✔ Is the definition correct? ✔ Would the definition help someone to learn?

❑ Yes, it is correct ❑ Yes, it is helpful

❑ No, it is wrong ❑ No, it is not helpful

❑ I am not sure ❑ I am not sure

I think this because:

_____

_____

_____

_____

_____

RS•C

b) **Molecule: Something that is formed by two atoms bonding together.**

✔ Is the definition correct?   ✔ Would the definition help someone to learn?

❑ Yes, it is correct   ❑ Yes, it is helpful

❑ No, it is wrong   ❑ No, it is not helpful

❑ I am not sure   ❑ I am not sure

I think this because:

_____

_____

_____

_____

_____

c) **Molecule: The smallest portion of a substance capable of existing independently and retaining the properties of the original substance.**

✔ Is the definition correct?   ✔ Would the definition help someone to learn?

❑ Yes, it is correct   ❑ Yes, it is helpful

❑ No, it is wrong   ❑ No, it is not helpful

❑ I am not sure   ❑ I am not sure

I think this because:

_____

_____

_____

_____

_____

d) **Molecule: Group of two or more atoms bonded together. A molecule of an element consists of one or more like atoms; a molecule of a compound consists of two or more different atoms bonded together.**

✔ Is the definition correct?   ✔ Would the definition help someone to learn?

❑ Yes, it is correct   ❑ Yes, it is helpful

❑ No, it is wrong   ❑ No, it is not helpful

❑ I am not sure   ❑ I am not sure

I think this because:

_____

_____

_____

_____

_____

RS•C

This page has been inentionally left blank.

RS•C

RS•C

Types of chemical reaction

Target level

This probe is primarily designed for students in the 14–16 age range who have been introduced to the range of reaction types included. It may also be useful for checking the prior knowledge of students on post-16 courses.

Topics

Types of chemical reaction: displacement, neutralisation, oxidation, reduction, thermal decomposition.

Rationale

Students in the 14–16 year age range are expected to learn to recognise certain common types of reaction. Students are likely to come across a number of reactions in their studies. Recognising that a reaction is of a certain type can help the student fit the reaction into one of a limited number of classes. (See also the materials on **Word equations** for the 11–14 age range.) These ideas are discussed in Chapter 9 of the Teachers' notes.

During piloting, it was found that some students found the exercise difficult, but students recognised that it was useful and helped clarify concepts. Teachers reported that the probe was quick and simple to use, led to valuable discussion, and provided useful revision - enabling students to recognise patterns and connections 'between seemingly disparate examples'.

Instructions

Each student requires a copy of the worksheet **Types of chemical reaction**.

In the pilot some students found it difficult to explain the reasons for their classifications. Teachers may wish to ask some groups of students to just undertake the classification. Students may then be asked to explain the correct responses when the teacher goes through the examples.

It should be emphasised that some reactions may be classified in several categories. (Note that oxidation and reduction are shown as separate classes.)

Resources

■ Student worksheet
  – Types of chemical reaction

Feedback for students

A suggested answer sheet for teachers is provided. Teachers may wish to go through the answers for the example on the first page of the worksheet before their students attempt the rest of the exercise.

RS•C

Types of chemical reaction – answers

Example

magnesium + steam → magnesium oxide + hydrogen

$Mg(s) + H_2O(g) → MgO(s) + H_2(g)$

This is an example of an **oxidation** (magnesium is oxidised) and therefore also of a **reduction** (water is reduced).

It can also be seen as magnesium displacing hydrogen from water to form the oxide.

1. nitrogen + hydrogen → ammonia

   $N_2(g) + 3H_2(g) → 2NH_3(g)$

   This binary synthesis reaction an example of a **reduction** (nitrogen is reduced) and therefore also of an **oxidation** (hydrogen is oxidised).

2. sodium hydroxide + nitric acid → sodium nitrate + water

   $NaOH(aq) + HNO_3(aq) → NaNO_3(aq) + H_2O(l)$

   This is an example of a **neutralisation** reaction between an acid (nitric acid) and an alkali (sodium hydroxide).

3. copper carbonate + sulfuric acid → copper sulfate + water + carbon dioxide

   $CuCO_3(s) + H_2SO_4(aq) → CuSO_4(aq) + H_2O(l) + CO_2(g)$

   This is an example of an acid-base reaction, but is not normally called a neutralisation.

4. sodium + water → sodium hydroxide + hydrogen

   $2Na(s) + 2H_2O(l) → 2NaOH(aq) + H_2(g)$

   This is an example of an **oxidation** (sodium is oxidised), and therefore also of a **reduction** (hydrogen from water is reduced).

   This could also be considered as an example of a **displacement** reaction, with sodium displacing hydrogen.

5. zinc + copper sulfate → zinc sulfate + copper

   $Zn(s) + CuSO_4(aq) → ZnSO_4(aq) + Cu(s)$

   This is an example of a **displacement** reaction - with zinc displacing copper from the salt.

   This is also an example of an oxidation (zinc is oxidised) and therefore also of a reduction (copper is reduced).

6. copper carbonate → copper oxide + carbon dioxide

   $CuCO_3(s) → CuO(s) + CO_2(g)$

   This reaction requires heating, and is an example of a **thermal decomposition**.

7. sodium bromide + chlorine → sodium chloride + bromine

   $2NaBr(aq) + Cl_2(aq) → 2NaCl(aq) + Br_2(aq)$

   This is an example of a **displacement** reaction - with chlorine displacing bromine in the salt.

   It is also an example of an **oxidation** (bromide is oxidised to bromine) and a reduction (chlorine is reduced to chloride).

8. copper oxide + carbon → copper + carbon dioxide
   $2CuO(s) + C(s) \rightarrow 2Cu(s) + CO_2(g)$

   This is an example of a **displacement** reaction, with carbon displacing copper from the oxide.

   This is also an **oxidation** process (carbon is oxidised) and therefore a **reduction** (copper is reduced).

9. methane + oxygen → carbon dioxide + steam

   This combustion reaction is an **oxidation** process (carbon is oxidised) and therefore also a reduction (oxygen is reduced). Some students may feel that hydrogen is also oxidised (as it 'gains' oxygen), although from a perspective of oxidation states hydrogen is unchanged.

10. zinc + hydrochloric acid → zinc chloride + hydrogen
    $Zn(s) + 2HCl(aq) \rightarrow ZnCl_2(aq) + H_2(g)$

    It may also be an considered an example of a **displacement** reaction with zinc displacing hydrogen.

11. sodium chloride → sodium + chlorine

    This reaction requires an input of energy, such as in electrolysis. It is an example of an **oxidation** process (chlorine is oxidised) and therefore also of a **reduction** (sodium is reduced).

    or sodium nitrate(V) → sodium nitrate(III) + oxygen

    $2NaNO_3(s) \rightarrow 2NaNO_2(s) + O_2(g)$

    This is also an example of a **reduction** process (nitrogen is reduced), and therefore also of an **oxidation** process (some of the oxygen in the nitrate is oxidised to elemental oxygen).

Types of chemical reaction

Scientists classify chemical reactions into different types - such as oxidation and neutralisation.

This exercises provides the equations for a number of chemical reactions.

For each reaction you are given a word equation, and an equation using chemical symbols:

magnesium + steam → magnesium oxide + hydrogen

$Mg(s) + H_2O(g) → MgO(s) + H_2(g)$

You should try to classify each of the examples given.

✔	type of reaction
❑	displacement
❑	neutralisation
❑	oxidation
❑	reduction
❑	thermal decomposition
❑	none of the above

For each reaction tick (✔) the box, or boxes, that describe the type of reaction. Some of the reactions may be examples of more than one type of reaction.

Some of the reactions may only occur when energy is provided (as heat, or as electrical energy), but this is not shown in the questions.

Tick (✔) 'none of the above' if the reaction does not seem to fit any of the suggestions.

Explain why you have classified the reaction the way you have.

RS•C

1. **nitrogen + hydrogen → ammonia**
    $N_2(g) + 3H_2(g) \rightarrow 2NH_3(g)$

    ✔ type of reaction   I made this classification because:

    ❑ displacement   _____

    ❑ neutralisation   _____

    ❑ oxidation   _____

    ❑ reduction   _____

    ❑ thermal decomposition   _____

    ❑ none of the above   _____

2. **sodium hydroxide + nitric acid → sodium nitrate + water**
    $NaOH(aq) + HNO_3(aq) \rightarrow NaNO_3(aq) + H_2O(l)$

    ✔ type of reaction   I made this classification because:

    ❑ displacement   _____

    ❑ neutralisation   _____

    ❑ oxidation   _____

    ❑ reduction   _____

    ❑ thermal decomposition   _____

    ❑ none of the above   _____

3. **copper carbonate + sulfuric acid → copper sulfate + water + carbon dioxide**
    $CuCO_3(s) + H_2SO_4(aq) \rightarrow CuSO_4(aq) + H_2O(l) + CO_2(g)$

    ✔ type of reaction   I made this classification because:

    ❑ displacement   _____

    ❑ neutralisation   _____

    ❑ oxidation   _____

    ❑ reduction   _____

    ❑ thermal decomposition   _____

    ❑ none of the above   _____

4. **sodium + water → sodium hydroxide + hydrogen**
   $2Na(s) + 2H_2O(l) → 2NaOH(aq) + H_2(g)$

✔ type of reaction   I made this classification because:

❑ displacement   _____

❑ neutralisation   _____

❑ oxidation   _____

❑ reduction   _____

❑ thermal decomposition   _____

❑ none of the above   _____

5. **zinc + copper sulfate → zinc sulfate + copper**
   $Zn(s) + CuSO_4(aq) → ZnSO_4(aq) + Cu(s)$

✔ type of reaction   I made this classification because:

❑ displacement   _____

❑ neutralisation   _____

❑ oxidation   _____

❑ reduction   _____

❑ thermal decomposition   _____

❑ none of the above   _____

6. **copper carbonate → copper oxide + carbon dioxide**
   $CuCO_3(s) → CuO(s) + CO_2(g)$

✔ type of reaction   I made this classification because:

❑ displacement   _____

❑ neutralisation   _____

❑ oxidation   _____

❑ reduction   _____

❑ thermal decomposition   _____

❑ none of the above   _____

RS•C

7. **sodium bromide + chlorine → sodium chloride + bromine**
   $2NaBr(aq) + Cl_2(aq) \rightarrow 2NaCl(aq) + Br_2(aq)$

   ✔ type of reaction   I made this classification because:

   ❑ displacement   _____

   ❑ neutralisation   _____

   ❑ oxidation   _____

   ❑ reduction   _____

   ❑ thermal decomposition   _____

   ❑ none of the above   _____

8. **copper oxide + carbon → copper + carbon dioxide**
   $2CuO(s) + C(s) \rightarrow 2Cu(s) + CO_2(g)$

   ✔ type of reaction   I made this classification because:

   ❑ displacement   _____

   ❑ neutralisation   _____

   ❑ oxidation   _____

   ❑ reduction   _____

   ❑ thermal decomposition   _____

   ❑ none of the above   _____

9. **methane + oxygen → carbon dioxide + steam**
   $CH_4(g) + 2O_2(g) \rightarrow CO_2(g) + 2H_2O(g)$

   ✔ type of reaction   I made this classification because:

   ❑ displacement   _____

   ❑ neutralisation   _____

   ❑ oxidation   _____

   ❑ reduction   _____

   ❑ thermal decomposition   _____

   ❑ none of the above   _____

10. **zinc + hydrochloric acid → zinc chloride + hydrogen**
$$Zn(s) + 2HCl(aq) \rightarrow ZnCl_2(aq) + H_2(g)$$

✔ type of reaction   I made this classification because:

❏ displacement   _____

❏ neutralisation   _____

❏ oxidation   _____

❏ reduction   _____

❏ thermal decomposition   _____

❏ none of the above   _____

11. **sodium chloride → sodium + chlorine**
$$2NaCl(l) \rightarrow 2Na(l) + Cl_2(g)$$

✔ type of reaction   I made this classification because:

❏ displacement   _____

❏ neutralisation   _____

❏ oxidation   _____

❏ reduction   _____

❏ thermal decomposition   _____

❏ none of the above   _____

12. **sodium nitrate → sodium nitrite + oxygen**
or
sodium nitrate(V) → sodium nitrate(III) + oxygen

$$2NaNO_3(s) \rightarrow 2NaNO_2(s) + O_2(g)$$

✔ type of reaction   I made this classification because:

❏ displacement   _____

❏ neutralisation   _____

❏ oxidation   _____

❏ reduction   _____

❏ thermal decomposition   _____

❏ none of the above   _____

RS•C

Revising the Periodic Table

Target level

This exercise is intended for students in the 14–16 year old age range, who have studied the Periodic Table. It may also be useful as a way of testing prior knowledge for students on post-16 courses.

Topic

The Periodic Table.

Rationale

The Periodic Table is one of the central organising themes in chemistry, and one of the areas where ideas about the macroscopic (*ie* properties of elements) and molecular (*eg* electronic configuration) properties are related (and often confused) – see Chapter 6 of the Teachers' notes.

This exercise is based on concept mapping - an important method for exploring students ideas, and a useful study and revision technique. (No prior experience of concept mapping is needed to undertake the exercise.) The exercise provides a concept map for the Periodic Table, with links labelled with numbers. The students are asked to supply (30) sentences to match the links. The exercise can be set as an individual task to elicit student knowledge, or as a group task to encourage discussion of ideas. Concept mapping is discussed in Chapter 3 of the Teachers' notes.

During piloting, it was felt that the exercise was thought provoking for students, and required them to think about concise definitions of terms. Some students reacted positively to the activity, although some found it boring. (See the comments about different activities appealing to students with different learning styles in Chapter 3 of the Teachers' notes.) The exercise was found to reveal areas where students were confused by what is meant by 'substance' in chemistry for example - see Chapter 6 of the Teachers' notes).

Instructions

Each student requires a copy of the worksheets **Revision map for the Periodic Table** and **The Periodic Table** – a sheet for student responses

Teachers may wish to encourage some students to add new concepts and links to the map as an extension activity.

Resources

Teachers who use the term 'outer shell electrons' rather than 'valence electrons' may wish to modify the worksheets.

■ Student worksheets
  – Revision map for the Periodic Table
  – The Periodic Table
  – Revision summary of the Periodic Table

RS•C

Feedback for students

A sample answer sheet for students (**Revision summary for the Periodic Table**) is provided. Teachers may wish to issue this when students have undertaken the exercise. Where students in the group have made creditable suggestions for additional concepts/links, then the teacher may wish to have students in the group add these to the answer sheet.

RS•C

Revision map for the Periodic Table

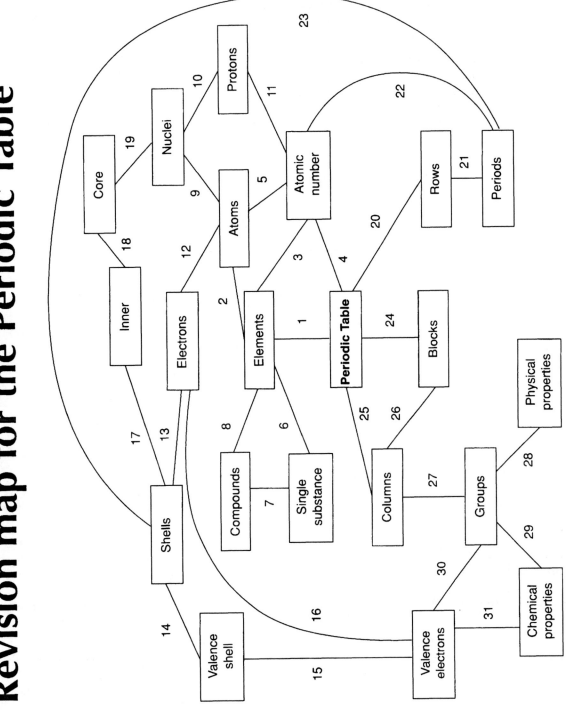

RS•C

The Periodic Table

You have been given a copy of a revision 'map' for the topic of the Periodic Table. The map is a diagram with boxes labelled with some words we use in chemistry, joined by numbered lines. Each of the lines suggests a relationship, which could be described in a sentence.

You are asked to think up sentences to show how the different words in the boxes are related. Try and be as accurate and precise as you can. Make sure you put each sentence next to the correct number in the spaces below. One of the sentences has been suggested for you, to get you started.

Fill in as many of the spaces as you can, **but do not worry** if you cannot complete them all.

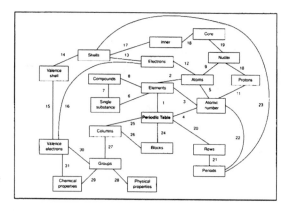

1.

2.

3.

4.

5.

6.   An element is a single chemical substance.

7.

8.

9.

10.

11.

12.

RS•C

13.

14.

15.

16.

17.

18.

19.

20.

21.

22.

23.

24.

25.

26.

27.

28.

29.

30.

31.

Revision summary for the Periodic Table

Some suggested sentences linking the words in the diagram.

1. The Periodic Table is a way of arranging what we know about the chemical elements.

2. Each element in the Periodic Table is a different type of atom.

3. Each element has a different atomic number.

4. The Periodic Table is arranged in atomic number order.

5. Each atom has an atomic number.

6. An element is a single chemical substance.

7. A compound is a single chemical substance.

8. Compounds are formed when two or more chemical elements react together.

9. An atom has a central nucleus.

10. A nucleus contains one or more protons.

11. The number of protons in the atomic nucleus gives the atomic number.

12. An atom contains one or more electrons.

13. The electrons are arranged in shells.

14. The outermost shell is called the valence shell.

15. Electrons in the valence shell are called valence electrons.

16. Each atom has one or more valence electrons.

17. Most atoms have one or more inner shells of electrons inside the valence shell.

18. The inner shells are part of the atomic core.

19. The nucleus is part of the atomic core.

20. The elements are arranged in rows in the Periodic Table.

21. The rows of elements are called periods.

22. The higher the atomic number of an element, the higher the period.

23. The number of shells of electrons determines the period an element is in.

24. The Periodic Table is arranged in several blocks.

25. The elements are arranged in columns in the Periodic Table.

26. Each block of the Periodic Table contains several columns that are next to each other.

27. The columns of elements are called groups.

28. Physical properties of elements often change in a pattern down the group.

29. Groups of elements usually have similar chemical properties.

30. Elements in a group have the same number of outer valence electrons.

31. An element's chemical properties are related to the number of outer valence electrons in its atoms.

RS•C

The melting temperature of carbon

Target level

This exercise is suitable for either 14–16 year olds who have studied bonding and structure and can calculate relative molecular mass, or for revision with post-16 students prior to meeting these topics at an advanced level. A knowledge of temperature in kelvin (K) is required.

Topics

The relationship between structure and properties; giant covalent structures.

Rationale

It is known that students may have difficulty in distinguishing between substances with simple molecules (where the solid is held together by intermolecular forces), and those with giant covalent lattices. This problem is compounded by the convention of giving carbon allotropes the formula C (*cf* He, Ne etc), and representing the relative molecular mass (M_r) accordingly. (The melting temperature and structure used are that of diamond – to avoid the complication of the anisotropy of graphite.)

This exercise is intended to make students think about this issue, and to realize that although diamond is commonly given the formula 'C' , this should not be taken to imply it forms monatomic molecules. These ideas are discussed in Chapter 7 of the Teachers' notes.

During piloting, teachers suggested that the exercise was 'useful in making pupils think about what a formula means' and 'useful to reinforce the difference between simple and giant covalent'. Some post-16 students thought the exercise 'clarified things', but it was found too basic by others at this level.

Some students will recall that carbon (diamond or graphite) has a high melting temperature. This should not diminish the effectiveness of the exercise, as long as students recognise the general pattern in the data provided, and think about why carbon is anomalous. Similarly, it is not important whether students are already familiar with '∞' as the symbol for infinity, as long as they recognise that the formula 'C' should not be taken to imply carbon is monatomic.

The kelvin (K) is the unit of thermodynamic temperature difference. It is defined by setting the thermodynamic temperature of the triple point of water at 273.16 K above absolute thermodynamic zero.

The 'degree Celsius' as a measure of temperature difference is no longer used. It is the same size as the kelvin. The temperature difference between the freezing and boiling points of water at normal pressure on the Celsius scale (0 °C and 100 °C respectively) is therefore 100 K, since 0 °C = 273 K and 100 °C = 373 K.

Instructions

First issue the student worksheet **Predicting the melting temperature of carbon** and when this is completed, then hand out the worksheet **Explaining the melting temperature of carbon**.

RS•C

Resources

■ Student worksheets
  – Predicting the melting temperature of carbon
  – Explaining the melting temperature of carbon

Feedback for students

A suggested answer sheet for teachers is provided.

RS•C

RS•C

The melting temperature of carbon – answers

Predicting the melting temperature of carbon

Predicting the melting temperature of carbon

1. $M_r(Ne) = 1 \times 20.2 = 20.2$

2. $M_r(Cl_2) = 2 \times 35.5 = 71.0$

3. $M_r(C) = 1 \times 12.0 = 12.0$

4. Students should recognise that in the examples given melting temperature increases with relative molecular mass.

5. The students should predict a melting temperature greater than 4K and less than 25K – and explain that they have selected an estimate to have a higher melting temperature than helium, and a lower melting temperature than neon.

(Experience from the pilot suggests that some students may provide quite precise estimates based on various algebraic manipulations of subsets of the data. It is suggested that such students could be asked to plot a graph, and comment on the reliability of their estimate.)

Explaining the melting temperature of carbon

The students should enter (in the box at the top of the page) the same predictions that they have given on the previous sheet.

If the prediction was in line with the data provided, (eg 10–20 K) then they should tick the box labelled 'a long way out'.

1. The diagrams should help explain the melting temperature of carbon because:

   unlike the other examples (neon, chlorine and sulfur) carbon does not comprise of separate (ie discrete) molecules, but a large (ie extensive) structure (lattice). Melting carbon requires breaking covalent bonds - not just overcoming intermolecular forces.

2. Giant covalent is meant to suggest:

   that the structure of carbon comprises of a very large number of units all interconnected through strong covalent bonds.

3. The symbol ∞ is meant to suggest either that the structure may be considered as effectively infinite or that a single carbon atom should not be considered to be a molecule of carbon.

Predicting the melting temperature of carbon

Introduction

Part of doing science involves spotting patterns, and making predictions which can be tested by experiments. In this exercise you will be asked to make a prediction based on information you will be given. It is not important whether your prediction is correct, as long as it is based on the information given.

Information about some elements

This section is to make sure you remember what is meant by the relative molecular mass of an element.

Many substances are said to be 'molecular' - they are comprised of vast numbers of separate identical particles. The name given to the tiny particles in these substances is molecules. Usually a molecule can be thought of as several atoms bonded together. In a few substances the molecules are single atoms. Often these substances are called atomic substances.

The formula of helium is He. It consists of single atoms. The relative atomic mass of helium, $A_r(He)$, is 4.0, and its relative molecular mass, $M_r(He)$, is also 4.0.

The formula of fluorine is F_2. It has molecules that can be thought of as two atoms bonded together. The relative atomic mass of fluorine, $A_r(F)$, is 19.0, and its relative molecular mass, $M_r(F_2)$, is therefore 38.0.

The formula of sulfur is S_8. It has molecules that can be thought of as eight atoms bonded in a ring arrangement. The relative atomic mass of sulfur, $A_r(S)$, is 32.1, and its relative molecular mass, $M_r(S_8)$, is therefore 256.8.

1. The formula of neon is Ne. It consists of single atoms. The relative atomic mass of neon, $A_r(Ne)$ is 20.2. What is the relative molecular mass of neon?

    $M_r(Ne) = $ _____

2. The formula of chlorine is Cl_2. It has molecules that that can be thought of as two atoms bonded together. The relative atomic mass of chlorine, $A_r(Cl)$, is 35.5. What is the relative molecular mass of chlorine?

    $M_r(Cl_2) = $ _____

3. The formula of carbon is C. The relative atomic mass of C is 12.0. What do you think the relative molecular mass of carbon will be?

    $M_r(C) = $ _____

The table below shows the melting temperatures (in Kelvin, K) of some elements.

Element	M_r	Melting temperature/K
Helium (He)	4.0	4
Carbon (C)	12.0	
Neon (Ne)	20.2	25
Fluorine (F$_2$)	38.0	53
Chlorine (Cl$_2$)	71.0	172
Bromine (Br$_2$)	159.8	266
Iodine (I$_2$)	253.8	387
Sulfur (S$_8$)	256.8	392

It has been suggested that there is a general relationship between the relative molecular mass of an element, and the temperature at which the solid element melts.

4. Can you see any relationship in the data given in the table? Describe any pattern you can see:

_____

_____

_____

_____

_____

The melting temperature of carbon is not given in the table.

5. Predict the (approximate) melting temperature for carbon? Make an estimate, and explain your reasons:

Solid carbon will melt at about _____

I think this because

_____

_____

_____

_____

Now ask your teacher for the second set of sheets.

Explaining the melting temperature of carbon

Write your prediction of the approximate melting temperature of carbon in the box below.

Prediction: about _____ K

This is your prediction based upon the data you were given. In science it is important to make predictions, because while testing the predictions we are also testing the ideas we use to explain things. Often when our predictions are wrong it helps us find better ways of understanding things.

Experiments show that solid carbon is difficult to melt and only changes into a liquid at a very high temperature. The melting temperature of carbon is 3823K. How close was your prediction? (tick one box)

❏ just about right   ❏ a little bit out   ❏ a long way out

On the separate sheet you will find four diagrams showing the particles in neon, chlorine, sulfur and carbon. It is not possible to draw accurate diagrams to show exactly what atoms and molecules are like. These picture are very simple models of how scientists sometimes think about these particles.

Look at the four diagrams, and use them to help you answer the following questions:

1. Do the diagrams help you understand why carbon has such a high melting temperature? (Explain your answer.)

    _____

    _____

    _____

    _____

2. Scientists sometimes describe carbon as having a giant covalent structure. What do you think is meant by giant covalent?

    _____

    _____

    _____

    _____

3. It is sometimes suggested that the symbol for the carbon macromolecule should be C_∞ rather than just C. What do you think the ∞ symbol is meant to show in C_∞?

    _____

    _____

    _____

    _____

The following diagrams show how scientists picture particles of neon (Ne), chlorine (Cl), sulfur (S) and carbon (C).

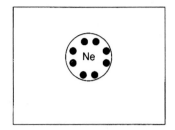

A neon atom, Ne

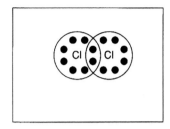

A chlorine molecule, Cl$_2$

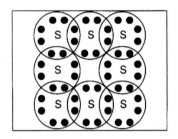

A sulfur molecule, S$_8$

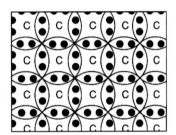

Part of the giant covalent
structure of carbon C or C$_\infty$

(There are many ways of drawing atoms and molecules - and we choose a type of diagram depending on which aspects we are interested in. For example, the diagrams above show the particles as flat. Molecules are not flat, but the shapes of the molecules were not important in this particular piece of work. At another time it might be important to show the shapes of the molecules.)

RS•C

This page has been intentionally left blank.

RS•C

Iron – a metal

Target level

This probe is designed for able students in the 14–16 age range who have been introduced to the properties of metals, and to the basic notion of metallic bonding, and for post-16 students.

Topics

Metallic structure and bonding; explaining properties using the particle model.

Rationale

Students often have considerable difficulty in using atomic/molecular-level models of matter to explain the properties of substances (see Chapter 6 of the Teachers' notes). It is also common for students to commence post-16 studies believing (a) that all materials have either covalent or ionic bonding, and (b) that ionic and metallic materials are molecular (see Chapter 8 of the Teachers' notes).

This exercise asks students to judge the veracity of twenty statements about one of the most familiar examples of a metal, iron.

During piloting, it was found that the items elicited student confusion between the properties of particles, and bulk properties. It was felt that the probe, and the feedback sheet, helped students clarify their ideas about the origins of the bulk properties of the metal. Teachers may be surprised at the number of students demonstrating the alternative conceptions targeted in this probe.

Instructions

It is worth reiterating to the students that the diagram shows just a small part of a slice through the lattice structure - and that the real structure is three-dimensional.

A blank answer sheet is provided offering the options 'true', 'false' and 'do not know' for each item. An alternative version of the answer sheet only includes the 'true' and 'false' options as some teachers prefer not to allow a 'do not know' option.

Resources

■ Student worksheets
 – Iron – true or false?
 – True or false? – response sheets (two versions)
 – Iron – answers

Feedback for students

A suggested feedback sheet is provided for students (**Iron – answers**) and teachers may wish to issue this after students have tackled the probe.

Iron usually has a body centred cubic lattice (with coordination number 8). Teachers may wish to point out that many common metals have a slightly different structure - being close packed with a coordination number of 12.

Iron – answers

Below you will find listed the 20 statements you were asked to think about. Following each is a brief comment suggesting whether or not the statement is true, and why.

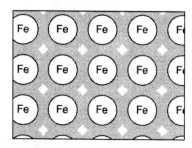

1. Iron has a type of bonding called metallic bonding. **TRUE**: iron is a metal, and all metals have a type of bonding called metallic bonding which is different from covalent and ionic bonding. In metallic bonding the outer shells of adjacent atoms overlap, and the outer shell electrons are free to move about through the lattice. The metal consists of metal cations and a balancing number of these 'free' electrons.

2. Iron atoms do not have a full outer shell of electrons, and this makes iron very reactive. **FALSE**: although an isolated iron atom has an electronic configuration of 2.8.14.2, the outer electrons are involved in the bonding in the metal. Iron is not very reactive, although it will slowly rust.

3. An iron atom is a silver-grey colour, and so iron metal is a silver-grey colour. **FALSE**: the colour of iron is a property of the arrangement of cations and electrons. A single atom of iron would not have a colour.

4. Iron can conduct electricity because some of the iron atoms can slip over their neighbours, and move through the solid. **FALSE**: the iron cations are normally fixed in their lattice positions. It is the electrons from the outer shells that are able to move about, allowing electrical current to flow through the metal.

5. Iron can be reshaped, without changing the shape of iron atoms. **TRUE**: metals can be worked into different shapes by hammering to force the cations to slip over each other. The cations change position, but not shape.

6. The reason iron rusts is that iron atoms will rust if exposed to damp air. **FALSE**: the rusting of iron is due to a chemical reaction between the iron and oxygen and water vapour in the air. During these reactions some of the iron cations and electrons become part of a new chemical compound (the rust), but the atoms themselves do not corrode.

7. In iron metal each atom is bonded to each of the other iron atoms surrounding it. **TRUE**: the iron atoms are packed together so that each iron cation is surrounded by eight others as if it is in the centre of a cube. The structure is held together by metallic bonding.

8. Iron conducts electricity because iron atoms are electrical conductors. **FALSE**: the metal conducts because some electrons are able to move through the metallic lattice structure. The individual atoms can not be considered to conduct. The outer electrons are only able to leave the cations because the outer electron shells overlap.

9. Iron is a solid because that is the natural state for metals. **FALSE**: the natural state depends on the temperature. Deep in the earth – where it is very hot – iron is a liquid. One metal called mercury is a liquid at room temperature.

10. A metal such as iron consists of positive metal ions, and negative electrons which move around the solid between the ions. **TRUE**: the iron structure contains iron cations surrounded by the fast moving electrons that would be in outer shells of separate iron atoms. (Sometimes this is called a 'sea of electrons'.)

11. An iron atom will reflect light, and so freshly polished iron shines. **FALSE**: polished metal will form a mirror because of the regular lattice of cations and the 'sea' of electrons. Individual iron atoms would not reflect light.

12. The reason that iron becomes a liquid when heated is because the bonds melt. **FALSE**: the metal melts when enough energy is provided to allow the cations to slip over each other. The bonds in the liquid metal are weaker than in a solid metal. If the liquid was heated until it boiled the bonds would break (but not 'melt').

13. Iron conducts electricity because it contains a 'sea' of electrons. **TRUE**: the electrons are able to move about, and will pass along the metal when it is connected to a battery.

14. The atoms in a metal such as iron are held together by ionic bonds. **FALSE**: the bonding in a metal is metallic bonding. This is different from ionic bonding as there are no anions (negative ions) present.

15. The reason iron conducts heat is because there is room between the atoms for hot air to move through the metal. **FALSE**: the iron cations are held close together by the metallic bonding, and there is no room for other atoms and molecules to get between them. Heat passes along the metal due to lattice vibrations and the movement of electrons.

16. The reason that iron is hard, is because iron atoms are hard. **FALSE**: hardness is a property of the metal due to the strong bonding holding the structure together. It is the arrangement of cations and free electrons which makes the metal hard.

17. In iron, there are molecules held together by magnetism. **FALSE**: there are no molecules in a metal - each cation is bonded to all those around it by the 'sea' of electrons, and those cations are bonded to others, and so on. Each cation in a metallic crystal is bonded (indirectly) to all the others.

18. If a metal such as iron is heated to a very high temperature it would become a gas. **TRUE**: if a solid metal is heated it will melt, and if heating is continued to a high enough temperature the liquid metal will boil.

19. Metals such as iron expand when heated because the atoms get bigger. **FALSE**: when the metal is heated the cations vibrate more, and move a little further apart.

20. Chemical bonds are needed to hold the atoms together in a metal such as iron, even though all of the atoms are of the same type. **TRUE**: the atoms would not remain joined together if there was no bonding between them. This is true for all solids whether the atoms are of one type (in an element) or several (in a compound).

Iron – true or false?

The statements below refer to the diagram of the structure of iron. The diagram shows part of a slice through the three dimensional structure.

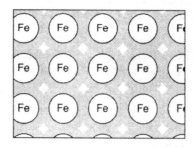

Please read each statement carefully, and decide whether it is correct or not.

1. Iron has a type of bonding called metallic bonding.

2. Iron atoms do not have a full outer shell of electrons, and this makes iron very reactive.

3. An iron atom is a silver-grey colour, and so iron metal is a silver-grey colour.

4. Iron can conduct electricity because some of the iron atoms can slip over their neighbours, and move through the solid.

5. Iron can be reshaped, without changing the shape of iron atoms.

6. The reason iron rusts is that iron atoms will rust if exposed to damp air.

7. In iron metal each atom is bonded to each of the other iron atoms surrounding it.

8. Iron conducts electricity because iron atoms are electrical conductors.

9. Iron is a solid because that is the natural state for metals.

10. A metal such as iron consists of positive metal ions, and negative electrons which move around the solid between the ions.

11. An iron atom will reflect light, and so freshly polished iron shines.

12. The reason that iron becomes a liquid when heated is because the bonds melt.

13. Iron conducts electricity because it contains a 'sea' of electrons.

14. The atoms in a metal such as iron are held together by ionic bonds.

15. The reason iron conducts heat is because there is room between the atoms for hot air to move through the metal.

16. The reason that iron is hard is because iron atoms are hard.

17. In iron there are molecules held together by magnetism.

18. If a metal such as iron is heated to a very high temperature it would become a gas.

19. Metals such as iron expand when heated because the atoms get bigger.

20. Chemical bonds are needed to hold the atoms together in a metal such as iron, even though all of the atoms are of the same type.

True or false? – response sheet

1.	True	False	1.
2.	True	False	2.
3.	True	False	3.
4.	True	False	4.
5.	True	False	5.
6.	True	False	6.
7.	True	False	7.
8.	True	False	8.
9.	True	False	9.
10.	True	False	10.
11.	True	False	11.
12.	True	False	12.
13.	True	False	13.
14.	True	False	14.
15.	True	False	15.
16.	True	False	16.
17.	True	False	17.
18.	True	False	18.
19.	True	False	19.
20.	True	False	20.

True or false? – response sheet

1.	True	Do not know	False	1.
2.	True	Do not know	False	2.
3.	True	Do not know	False	3.
4.	True	Do not know	False	4.
5.	True	Do not know	False	5.
6.	True	Do not know	False	6.
7.	True	Do not know	False	7.
8.	True	Do not know	False	8.
9.	True	Do not know	False	9.
10.	True	Do not know	False	10.
11.	True	Do not know	False	11.
12.	True	Do not know	False	12.
13.	True	Do not know	False	13.
14.	True	Do not know	False	14.
15.	True	Do not know	False	15.
16.	True	Do not know	False	16.
17.	True	Do not know	False	17.
18.	True	Do not know	False	18.
19.	True	Do not know	False	19.
20.	True	Do not know	False	20.

RS•C

RS•C

Ionic bonding

Target level

This probe may be used with students in the 14–16 age range who have studied chemical bonding, or with post-16 chemistry students .

Topic

Ionic bonding.

Rationale

This probe is designed to elicit alternative conceptions about ionic bonding that have been found to be commonplace among students. In particular, the probe explores the extent to which students see ionic bonding in sodium chloride as a molecular phenomena, with discrete NaCl ion pairs which are internally ionically bonded, but attracted to each other by weaker forces. These ideas are discussed in Chapter 8 of the Teachers' notes.

The probe presented here is an edited version of the draft which was piloted, and which was considered to be too long. When the probe was tried out in schools and colleges it was considered 'excellent'. Teachers reported that it led to 'interesting results' and 'lively debate' which brought out many misconceptions. It was considered to help clarify points and to provoke students into asking questions and thinking about the topic. It was considered by students to be 'useful', 'good for revision' and 'interesting' - and some even found it enjoyable!

Instructions

It is worth reiterating to the students that the diagram shows just a small part of a slice through the lattice structure - and that the real structure is three-dimensional.

A blank answer sheet is provided offering the options 'true', 'false' and 'do not know' for each item. An alternative version of the answer sheet only includes the 'true' and 'false' options as some teachers prefer not to allow a 'do not know' option.

Resources

- ■ Student worksheets
  – Ionic bonding - true or false?
  – True or false? - response sheets (two versions)
  – Ionic bonding - answers

Feedback for students

A suggested feedback sheet is provided for students (**Ionic bonding – answers**) and teachers may wish to issue this after students have tackled the probe.

Ionic bonding – answers

Below you will find listed the 20 statements you were asked to think about. Following each is a brief comment suggesting whether or not the statement is true, and why.

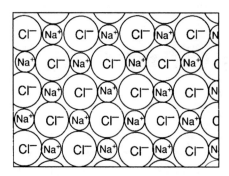

1. A positive ion will be attracted to any negative ion. **TRUE**: any positively charged object will be attracted to any negatively charged object. It does not matter how the objects acquired the charge, the attraction only depends on the amount of charge, and the distance between the two charged objects.

2. A sodium ion is only bonded to the chloride ion it donated its electron to. **FALSE**: each positive sodium ion is bonded to each of the neighbouring negative chloride counter-ions. It is irrelevant how the ions came to be charged.

3. A sodium atom can only form one ionic bond, because it only has one electron in its outer shell to donate. **FALSE**: a sodium ion can strongly bond to as many chloride ions as can effectively pack around it in the regular crystal lattice. In NaCl there will be six chloride ions strongly bonded to each sodium ion.

4. The reason a bond is formed between chloride ions and sodium ions is because an electron has been transferred between them. **FALSE**: The reason a bond is formed between chloride ions and sodium ions is because they have opposite electrostatic charges - negative and positive.

5. In the diagram a chloride ion is attracted to one sodium ion by a bond and is attracted to up to three other sodium ions just by forces. **FALSE**: In the diagram each chloride ion is attracted to up to four sodium ions by a bond that is an electrostatic force. (There would also be a fifth sodium ion above the chlorine ion and one more below - but these are not shown in the diagram.)

6. In the diagram each molecule of sodium chloride contains one sodium ion and one chloride ion. **FALSE**: there are no molecules in sodium chloride, just ions. A molecule comprises a group of atoms strongly bound together, and only weakly bonded (if at all) to other molecules. In sodium chloride each ion is strongly bonded to each of its six nearest neighbours.

7. An ionic bond is the attraction between a positive ion and a negative ion. **TRUE**.

8. A positive ion can be bonded to any neighbouring negative ion, if it is close enough. **TRUE**. The bond is just the attraction between the oppositely charged ions. If the ions are close together this force will be a strong bond.

9. A negative ion will be attracted to any positive ion. **TRUE**: any negatively charged object will be attracted to any positively charged object. It does not matter how the objects acquired the charge, the attraction only depends on the amount of charge, and the distance between the two charged objects.

10. It is not possible to know where the ionic bonds are, unless you know which chloride ions accepted electrons from which sodium ions. **FALSE**: as the bonding is just the attraction between ions, there will be a bond between any adjacent oppositely charged ions.

11. A chloride ion is only bonded to the sodium ion it accepted an electron from. **FALSE**: each negative chloride ion is bonded to each of the neighbouring positive sodium counter-ions. It is irrelevant how the ions came to be charged.

12. A chlorine atom can only form one strong ionic bond, because it can only accept one more electron into its outer shell. **FALSE**: a chloride ion can strongly bond to as many sodium ions as can effectively pack around it in the regular crystal lattice. In NaCl there will be six sodium ions strongly bonded to each chloride ion.

13. There is a bond between the ions in each molecule, but no bonds between the molecules. **FALSE**: there are no molecules in sodium chloride, but a continuous network of bonds throughout the lattice.

14. A negative ion can only be attracted to one positive ion. **FALSE**: there is no limit to the number of positive ions that a negative ion can be attracted to (although there is a limit to how many can cluster around it).

15. The reason a bond is formed between chloride ions and sodium ions is because they have opposite charges. **TRUE**: the opposite charges attract them together, and this force of attraction is the ionic bond.

16. In the diagram a sodium ion is attracted to one chloride ion by a bond and is attracted to three other chloride ions just by forces. **FALSE**: In the diagram each sodium ion is attracted to up to four chloride ions by a bond that is an electrostatic force. (There would also be a fifth chloride ion above the sodium ion and one more (a sixth) below - but these are not shown in the diagram.)

17. A positive ion can only be attracted to one negative ion. **FALSE**: there is no limit to the number of positive ions that a negative ion can be attracted to (although there is a limit to how many can cluster around it).

18. An ionic bond is when one atom donates an electron to another atom, so that they both have full outer shells. **FALSE**: an ionic bond is the electrostatic force which holds two oppositely charged ions together. The ions could have become charged by electron transfer, but usually the ions were charged long before they came into contact. The bond is no stronger in the few cases where an electron has transferred between two atoms to give the ions that have become bonded.

19. A negative ion can be bonded to any neighbouring positive ion, if it is close enough. **TRUE**. The bond is just the attraction between the oppositely charged ions. If the ions are close together this force will be a strong bond.

20. There are no molecules shown in the diagram. **TRUE**: A molecule comprises a group atoms strongly bound together, and only weakly bonded (if at all) to other molecules. In sodium chloride each ion is strongly bonded to each of its six nearest neighbours.

Ionic bonding – true or false?

The statements below refer to the diagram of the structure of sodium chloride. **The diagram shows part of a slice through the three dimensional crystal structure.**

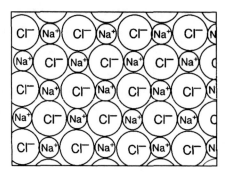

Please read each statement carefully, and decide whether it is correct or not.

1. A positive ion will be attracted to any negative ion.

2. A sodium ion is only bonded to the chloride ion it donated its electron to.

3. A sodium atom can only form one ionic bond, because it only has one electron in its outer shell to donate.

4. The reason a bond is formed between chloride ions and sodium ions is because an electron has been transferred between them.

5. In the diagram a chloride ion is attracted to one sodium ion by a bond and is attracted to other sodium ions just by forces.

6. In the diagram each molecule of sodium chloride contains one sodium ion and one chloride ion.

7. An ionic bond is the attraction between a positive ion and a negative ion.

8. A positive ion can be bonded to any neighbouring negative ions, if it is close enough.

9. A negative ion can be attracted to any positive ion.

10. It is not possible to point to where the ionic bonds are, unless you know which chloride ions accepted electrons from which sodium ions.

11. A chloride ion is only bonded to the sodium ion it accepted an electron from.

12. A chlorine atom can only form one ionic bond, because it can only accept one more electron into its outer shell.

13. There is a bond between the ions in each molecule, but no bonds between the molecules.

14. A negative ion can only be attracted to one positive ion.

15. The reason a bond is formed between chloride ions and sodium ions is because they have opposite charges.

16. In the diagram a sodium ion is attracted to one chloride ion by a bond and is attracted to other chloride ions just by forces.

17. A positive ion can only be attracted to one negative ion.

18. An ionic bond is when one atom donates an electron to another atom, so that they both have full outer shells.

19. A negative ion can be bonded to any neighbouring positive ions if it is close enough.

20. There are no molecules shown in the diagram.

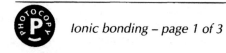

True or false? – response sheet

1.	True	False	1.
2.	True	False	2.
3.	True	False	3.
4.	True	False	4.
5.	True	False	5.
6.	True	False	6.
7.	True	False	7.
8.	True	False	8.
9.	True	False	9.
10.	True	False	10.
11.	True	False	11.
12.	True	False	12.
13.	True	False	13.
14.	True	False	14.
15.	True	False	15.
16.	True	False	16.
17.	True	False	17.
18.	True	False	18.
19.	True	False	19.
20.	True	False	20.

True or false? – response sheet

1.	True	Do not know	False	1.
2.	True	Do not know	False	2.
3.	True	Do not know	False	3.
4.	True	Do not know	False	4.
5.	True	Do not know	False	5.
6.	True	Do not know	False	6.
7.	True	Do not know	False	7.
8.	True	Do not know	False	8.
9.	True	Do not know	False	9.
10.	True	Do not know	False	10.
11.	True	Do not know	False	11.
12.	True	Do not know	False	12.
13.	True	Do not know	False	13.
14.	True	Do not know	False	14.
15.	True	Do not know	False	15.
16.	True	Do not know	False	16.
17.	True	Do not know	False	17.
18.	True	Do not know	False	18.
19.	True	Do not know	False	19.
20.	True	Do not know	False	20.

RS•C

RS•C

Precipitation

Target level

These materials are intended for 14–16 year students who are familiar with precipitation reactions. They may also be useful for diagnosing alternative conceptions of post-16 students.

Topics

Precipitation (double decomposition) reactions; solubility; formation of ionic bonding.

Rationale

These materials were inspired by the realisation that some post-16 students were unable to explain what was happening at the level of particles when precipitation occurs. This seems to be related to research findings that suggest students may not understand what happens during dissolving, and often have alternative conceptions of the nature of ionic bonding. These ideas are discussed in the Teachers' notes. Student problems with using molecular models are considered in Chapter 6; student ideas about ionic bonding are discussed in Chapter 8; and student ideas about the precipitation process are included in Chapter 9.

The materials comprise a diagnostic probe (on the precipitation of silver chloride), a follow-up study activity and a variation on the probe (on the precipitation of lead iodide). If the materials are used after teaching, then the silver chloride probe may be used to diagnose which students would benefit from remedial action. The study activity is suitable for setting as a private study ('homework') task. The second probe may be used to see if the exercise has helped students clarify their ideas.

During piloting, teachers found that some students were confusing atoms, ions, molecules and electrons, and that the probe revealed uncertainties about the nature of ionic substances: 'they had an idea of ionic bonding that was anchored to the idea of electron transfer, which led them along a 'diatomic molecule' type approach, thinking Ag^+ was only bonded to one Cl^-'.

Teachers found the study activity valuable, but it was considered a little repetitive for some students. The diagrams were considered to be helpful. Students commented that the exercise helped them imagine what was happening during the reaction. Some found it easy, and some enjoyed colouring the key.

Details of the DARTs activity can be found in Chapter 5 of the Teachers' notes.

Instructions

The worksheet **A reaction to form silver chloride** may be used as a diagnostic probe independently of the other materials. If students are unsure of the word 'particular', they should be asked to name the atoms/molecule/ions present.

The worksheet **A precipitation reaction** provides a simple exercise taking students through the reaction described in the probe. It could be made more difficult by deleting the initial letter of the responses.

The worksheet **A reaction to form lead iodide** may be used as a post-test after the study task is completed.

RS•C

Resources

■ Student worksheets
   – A reaction to form silver chloride
   – A precipitation reaction
   – A reaction to form lead iodide

■ Coloured pencils

Feedback for students

A suggested answer sheet for each of the three activities is provided for teachers.

RS•C

Precipitation – answers

A reaction to form silver chloride

1. H_2O molecules, Na^+, Cl^- ions (also allow H^+ or H_3O^+ and OH^- ions, as long as H_2O molecules given).

2. H_2O molecules, Ag^+, NO_3^- ions (also allow H^+ or H_3O^+ and OH^- ions, as long as H_2O molecules given).

3. H_2O molecules, Na^+, NO_3^- ions (also allow H^+ or H_3O^+ and OH^- ions, as long as H_2O molecules given).

   NB unless care is taken with reacting quantities the final solution may also contain either silver ions or chloride ions - but not both.

4. The important point is that the silver ions and chloride ions are already present in the mixture, and the reaction involves the electrical attraction causing the ions to clump together and form crystals.

   (Answers about ion formation through electron transfer are wrong.)

5. Students (especially those in the 14–16 age range) will not be expected to know about the precise crystal structure of silver chloride. Answers which suggest that the number of bonded ions depends upon the number of neighbours should be considered correct. For example, students may infer from the diagram that each ion is bonded to 4 or 6 others.

   Answers that are based on the charges on the ions (*ie* each silver ion is bonded to one chloride ion) are wrong, and may be related to irrelevant and inappropriate arguments about electron transfer.

A precipitation reaction

In this DART type activity (see Chapter 5 of the Teachers' notes) students have to complete missing words from the initial letters. The 'missing' words are shown here as bold text:

Sodium **chloride** is an ionic solid. Sodium ions (Na^+) and chloride ions (Cl^-) are bonded together by the **electrical** attraction between the positive and negative ions. Each **ion** is attracted to each of those counter ions surrounding it. This type of chemical **bonding** is called ionic bonding.

Sodium chloride dissolves in water. **Water** is a liquid containing water molecules. The **molecules** move around quickly, bumping into each other. In these **collisions** the water molecules bounce off one another. Many ionic solids will **dissolve** in water.

When the sodium chloride dissolves it forms a **solution**. The solution contains the water molecules, and the **sodium** ions and the **chloride** ions from the sodium chloride. The fast **moving** water molecules constantly collide with the ions, and crowd around ('solvate') them, so that the **ions** can not stick together.

Silver nitrate is an **ionic** solid. Silver ions (Ag^+) and **nitrate** ions (NO_3^-) are bonded together by the electrical **attraction** between the **positive** and negative ions. Each ion is attracted to each of those surrounding it. This type of **chemical** bonding is called ionic bonding.

When the silver nitrate **dissolves** it forms a solution. The solution contains the water **molecules**, and the silver ions and the nitrate **ions** from the silver nitrate. The **fast** moving water molecules constantly **collide** with the ions, and crowd around ('**solvate**') them, so that the ions can not stick together.

RS•C

In the mixture there would be:

water molecules, sodium cations, chloride anions, silver cations and nitrate anions

When the two solutions are **mixed** together the new mixture contains water **molecules**, sodium ions, **silver** ions, chloride ions and nitrate **ions**.

When silver ions collide with **chloride** ions they sometimes stick together. The **attraction** between these ions is so strong that collisions with the fast moving **water** molecules do not stop them bonding together.

The silver ions and **chloride** ions soon form into large enough crystals to **precipitate** out from the solution. The solid precipitate of silver chloride sinks to the **bottom** of the mixture.

In the **precipitate** silver ions (Ag^+) and chloride ions (Cl^-) are **bonded** together by the electrical attraction between the positive and negative **ions**. Each ion is attracted to **each** of those surrounding it. Silver chloride is an example of a compound with ionic bonding which does not dissolve in water (it is insoluble in **water**).

The liquid above the precipitate contains the water **molecules**, sodium ions and nitrate ions. This is a solution of sodium **nitrate**. The fast moving water molecules constantly collide with the ions, and crowd around ('solvate') them, so that the ions can not **stick** together.

The real change during the reaction is that **solvated** silver ions and solvated **chloride** ions form solid **silver** chloride:

The sodium ion and the nitrate ion (in the box at the bottom of the page) should be labelled as 'spectator ions'.

Summary sheet

Four different colours should be used to shade the four ions in the key (at foot of page). The ions in the other 5 figures should be coloured using the same colour code to show that the ions have effectively 'swapped partners'.

(It may be useful to make an overhead transparency of this page, and colour the ions. In this case the teacher may wish to set the colour code for the different ions – eg Na^+ blue, Ag^+ green, Cl^- yellow, NO_3^- red – so that each student can readily check their summary sheet against the version displayed on the overhead projector).

A reaction to form lead iodide

1. H_2O molecules, K^+, I^- ions (also allow H^+ or H_3O^+ ions and OH^- ions, as long as H_2O molecule is given).

2. H_2O molecules, Pb^{2+}, NO_3^- ions (also allow H^+ or H_3O^+ ions and OH^- ions, as long as H_2O molecule is given).

3. H_2O molecules, K^+, NO_3^- ions (also allow H^+ or H_3O^+ ions and OH^- ions, as long as H_2O molecule is given).

4. The lead ions and nitrate ions are already present in the mixture, and the reaction involves the electrical attraction causing the ions to clump together and form crystals.

5. Any answer that suggests that the number of bonded ions depends on the number of neighbours should be marked correct. Answers that are based on the charges of the ions are wrong.

A reaction to form silver chloride

Silver chloride is an ionic solid. It can be prepared by reacting silver nitrate solution and sodium chloride solution. The diagrams represent the types of particles present in some of the substances involved in the reaction.

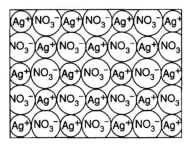

Silver nitrate (solid)

The diagram above shows the particles in solid silver nitrate. The particles in silver nitrate are silver ions (Ag^+) and nitrate ions (NO_3^-).

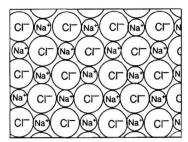

Sodium chloride (solid)

The diagram above shows the particles in solid sodium chloride. The particles in sodium chloride are sodium ions (Na^+) and chloride ions (Cl^-).

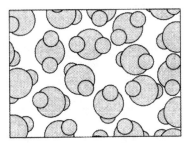

Water (liquid)

The diagram above shows the particles in water. Water is a liquid. The particles are water molecules.

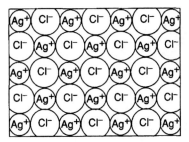

Silver chloride (solid)

The diagram above shows the particles in solid silver chloride. The particles in silver chloride are silver ions (Ag^+) and chloride ions (Cl^-).

A reaction to form silver chloride

The diagrams (on the other sheet) may help you with these questions.

1. Sodium chloride dissolves in water to give sodium chloride solution. What particles (such as particular atoms, molecules, ions) do you think are present in sodium chloride solution?

    _____

    _____

2. Silver nitrate dissolves in water to give silver nitrate solution. What particles (such as particular atoms, molecules, ions) do you think are present in silver nitrate solution?

    _____

    _____

    When sodium chloride solution is mixed with silver nitrate solution a white solid forms. The following reaction take place:

    sodium chloride(aq) + silver nitrate(aq) → sodium nitrate(aq) + silver chloride(s)

    The solid that is formed is silver chloride. The solid can be separated from the liquid by filtration. The liquid that is left after filtration contains sodium nitrate.

3. What particles (such as particular atoms, molecules, ions) do you think are present in the liquid after it is filtered?

    _____

    _____

4. What do you think happens to the particles in the mixture when the ionic bonds form in the silver chloride in this reaction?

    _____

    _____

    _____

    _____

5. How many chloride ions do you think are bonded to each silver ion in silver chloride? (Give the reason for your answer, if you can.)

    _____

    _____

    _____

A reaction for form silver chloride – page 2 of 2

RS•C

A precipitation reaction

This exercise is about what happens during a precipitation reaction. When solutions of sodium chloride and silver nitrate are mixed, then a white solid (silver chloride) forms:

sodium chloride(aq) + silver nitrate(aq) → sodium nitrate(aq) + silver chloride(s)

Fill in the gaps below. Use the diagrams to help you.

Sodium c_____ is an ionic solid. Sodium ions (Na^+) and chloride ions (Cl^-) are bonded together by the e_____ attraction between the positive and negative ions. Each i_____ is attracted to each of those counter ions surrounding it. This type of chemical b_____ is called ionic bonding.

Sodium chloride dissolves in water. W_____ is a liquid containing water molecules. The m_____ move around quickly, bumping into each other. In these c_____ the water molecules bounce off one another. Many ionic solids will d_____ in water.

When the sodium chloride dissolves it forms a s_____. The solution contains the water molecules, and the s_____ ions and the c_____ ions from the sodium chloride. The fast m_____ water molecules constantly collide with the ions, and crowd around ('solvate') them, so that the i_____ can not stick together.

Silver nitrate is an i_____ solid. Silver ions (Ag^+) and n_____ ions (NO_3^-) are bonded together by the electrical a_____ between the p_____ and negative ions. Each ion is attracted to each of those surrounding it. This type of c_____ bonding is called ionic bonding.

When the silver nitrate d_____ it forms a solution. The solution contains the water m_____, and the silver ions and the nitrate i_____ from the silver nitrate. The f_____ moving water molecules constantly c_____ with the ions, and crowd around ('s_____') them, so that the ions can not stick together.

What do you think would be in the mixture if the two solutions below were to be poured into the same beaker?

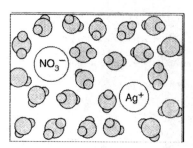

+

_____

_____

_____

_____

When the two solutions are m_____ together the

new mixture contains water m_____, sodium ions,

s_____ ions, chloride ions and nitrate i_____.

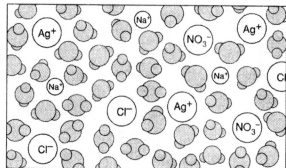

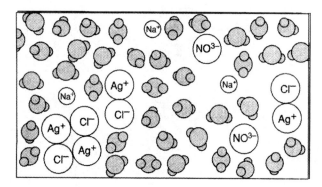

When silver ions collide with c_____ ions

they sometimes stick together. The a_____

between these ions is so strong that collisions with the

fast moving w_____ molecules do not stop

them bonding together.

The silver ions and c_____ ions soon form into

large enough crystals to p_____ out from the

solution. The solid precipitate of silver chloride sinks to the

b_____ of the mixture.

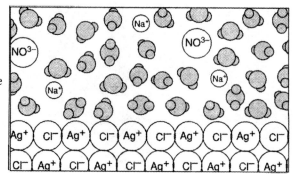

In the p_____ silver ions (Ag^+) and chloride ions (Cl^-) are b_____ together by the electrical attraction between the positive and negative i_____. Each ion is attracted to e_____ of those surrounding it. Silver chloride is an example of a compound with ionic bonding which does not dissolve in water (it is insoluble in w_____).

The liquid above the precipitate contains the water m_____, sodium ions and nitrate ions. This is a solution of sodium n_____. The fast moving water molecules constantly collide with the ions, and crowd around ('solvate') them, so that the ions can not s_____ together.

The real change during the reaction is that s_____ silver ions and solvated c_____ ions form solid s_____ chloride:

$$Cl^-(aq) + Ag^+(aq) \rightarrow AgCl(s)$$

The sodium ions and nitrate ions are sometimes called 'spectator' ions because they are not directly involved in forming the product. Label the spectator ions in the box below:

Ag^+ Cl^- Na^+ NO_3^-

Summary sheet

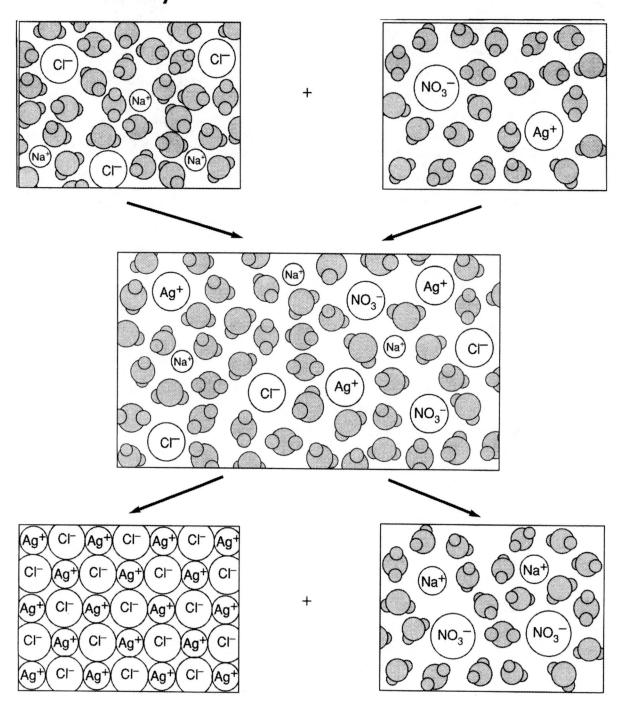

The change that has taken place in this reaction is more obvious if you use a colour key for the ions.

Select four different colours (*eg* blue, green, red and yellow), and colour each different type of ion in a different colour. Use the box below as a key.

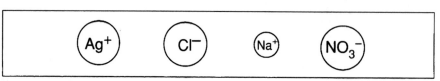

RS•C

A reaction to form lead iodide

Lead iodide is an ionic solid. It can be prepared by reacting lead nitrate solution and potassium iodide solution.

1. Potassium iodide dissolves in water to give potassium iodide solution. What particles (such as particular atoms, molecules, ions) do you think are present in potassium iodide solution?

   _____

   _____

2. Lead nitrate dissolves in water to give lead nitrate solution. What particles (such as particular atoms, molecules, ions) do you think are present in lead nitrate solution?

   _____

   _____

   When potassium iodide solution is mixed with lead nitrate solution a yellow solid forms. The following reaction take place:

   potassium iodide(aq) + lead nitrate(aq) → potassium nitrate(aq) + lead iodide(s)

   The solid that is formed is lead iodide. The solid can be separated from the liquid by filtration. The liquid that is left after filtration contains potassium nitrate.

3. What particles (such as particular atoms, molecules, ions) do you think are present in the liquid after it is filtered?

   _____

   _____

4. What do you think happens to the particles in the mixture, when the ionic bond forms in the lead iodide in this reaction?

   _____

   _____

   _____

   _____

5. How many iodide ions do you think might be bonded to each lead ion in lead iodide? (Give the reason for your answer, if you can.)

   _____

   _____

   _____

   _____

RS•C

This page has been intentionally left blank.

RS•C

RS•C

Spot the bonding

Target level

This probe is primarily intended for students undertaking or having completed post-16 chemistry courses. It may also be used with 14–16 year old students, to see if they can identify examples of the more limited range of bond types met at this level.

Topics

Chemical bonding (including: ionic, covalent, metallic, polar, hydrogen, dipole-dipole, van der Waals, solvation, dative, double, delocalised).

Rationale

Research suggests that students commonly focus on covalent and ionic bonding, and often fail to spot, or may down-play the importance of, other types of bonding. These ideas are discussed in Chapter 8 of the Teachers' notes. This probe provides a relatively quick way of auditing students' awareness of different bond types. (The probe **Interactions** will provide a means of exploring students' more detailed understanding of the same topic.)

A variety of types of diagram are used in this probe, as it is important for students to be able to interpret and use various ways of representing chemical species (see Chapter 6 of the Teachers' notes).

During piloting, teachers found this a 'very clear and very straight forward' probe, and a useful exercise for revising bonding and focusing and initiating discussion. Some students who had been taught about the types of intermolecular bonding were found to be confused about when different types of bonding would be found.

Instructions

It may be useful to point out to students that some of the diagrams refer to individual atoms or molecules, whilst others show the some of the particles in named substances. Students should therefore pay close attention to the labels under the figures.

Resources

■ Student worksheet
– Spot the bonding

Feedback for students

A suggested answer sheet is provided for teachers.

RS•C

Spot the bonding – answers

The following answers are suitable for students who have studied bonding at post-16 level. Where the **Spot the bonding** probe is used with students at an earlier stage, then they should not be expected to provide the full range of responses.

1. Sodium chloride lattice: ionic

2. Diamond lattice: covalent

3. Benzene molecule: covalent, delocalised

4. Copper lattice: metallic

5. Hydrogen fluoride molecule: covalent, polar

6. Liquid water: covalent, polar; hydrogen, covalent, van der Waals forces, dipole-dipole forces

7. Fluorine molecule: covalent

8. Sodium nitrate solution: covalent, polar; hydrogen, dipole-dipole, van der Waals forces, solvent-solute interactions

9. Oxygen gas: covalent (double/sigma + pi), van der Waals forces

10. Sulfur molecule: covalent

11. Sodium atom: no chemical bonding (although intra-atomic forces of similar nature)

12. Aluminium chloride dimer: polar, including dative (coordinate) covalent

13. Carbon dioxide molecule: covalent, polar (double/sigma + pi)

14. Ethanoic acid dimer: covalent, polar, hydrogen

15. Iodine lattice: covalent, van der Waals forces

16. Ammonia molecule: covalent, polar

17. Magnesium oxide lattice: ionic

18. Liquid hydrogen chloride: covalent, polar, van der Waals forces

Notes:

a) Where a bond has significant polarity, it could be described as polar rather than covalent (or polar covalent.)

b) The term van der Waals forces has been used for induced dipole-dipole forces.

c) Students may forget to mention van der Waals forces in cases where they recognise hydrogen-bonds are present (*ie* items 6, 8 and 18).

d) The presence of some covalent character in the magnesium oxide lattice may be spotted by some students.

Spot the bonding

This exercise comprises of a set of diagrams showing a range of chemical species and systems. For each diagram: either write the name or names of the type or types of bonding present, or write none (if there is no chemical bonding) or do not know if you are unsure.

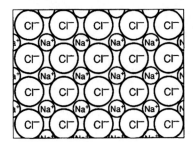

Sodium chloride lattice

1. _____

_____

_____

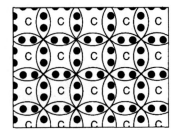

Diamond lattice

2. _____

_____

_____

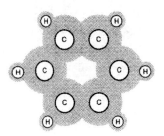

Benzene molecule

3. _____

_____

_____

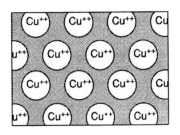

Copper metal lattice

4. _____

_____

_____

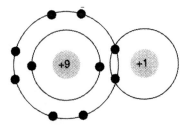

Hydrogen fluoride molecule

Hydrogen fluoride molecule

5. _____

_____

_____

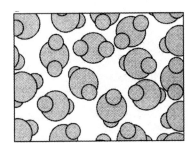

Liquid water

6. _____

_____

_____

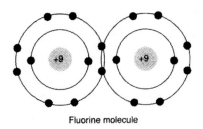

Fluorine molecule

Fluorine molecule

7. _____

_____

_____

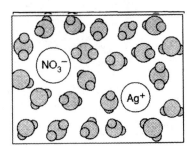

Sodium nitrate solution

8. _____

_____

_____

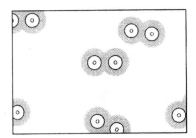

Oxygen gas

9. _____

_____

_____

RS•C

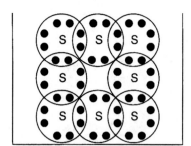

Sulfur molecule

10. _____

_____

_____

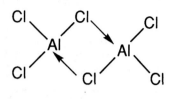

Sodium atom

11. _____

_____

_____

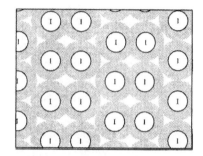

Aluminium chloride dimer

12. _____

_____

_____

$$O = C = O$$

Carbon dioxide molecule

13. _____

_____

_____

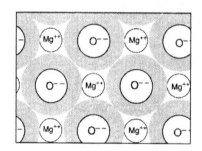

Ethanoic acid dimer

14. _____

_____

_____

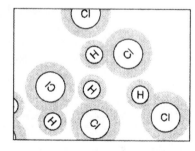

Iodine lattice

15. _____

_____

_____

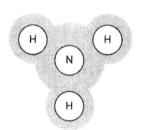

Ammonia molecule

16. _____

_____

_____

Magnesium oxide lattice

17. _____

_____

_____

Liquid hydrogen chloride

18. _____

_____

_____

RS•C

Hydrogen fluoride

Target level

This is intended primarily for use with groups of post-16 students. It may be used before teaching about energetics to elicit prior conceptions, or after teaching to see if the intended learning has taken place. The probe may also be used to elicit the ideas of students at earlier stages (*ie* 14–16 years olds), although students at this level may not have been taught a scientifically valid explanation for why reactions occur.

Topics

Driving force of chemical reactions: why reactions occur.

Rationale

Research suggests that many students believe that chemical reactions occur to enable atoms to obtain full outer electron shells, or 'octets', and that this belief is so strong that they tend to offer this explanation even when they have been taught more appropriate ideas.

These ideas are discussed in Chapters 9 and 10 of the Teachers' notes.

During piloting, it was reported that some students 'enjoyed' the task, and others found it 'demanding' and 'very challenging', but 'good revision' and 'useful in clarifying ideas'.

Teachers found their students' responses 'thought provoking'. One teacher described how 'students seemed to have searched for the first idea that they come across and tried to make it fit'. Another was 'disappointed' that the 17–18 year old students did not think about the thermodynamics of the question, despite having just studied the topic. Another teacher suggested that the probe showed 'clearly how wedded [the students in the group] are to 'happy' atoms'. The probe was described as 'a useful exercise – if depressing for the teacher'.

Resources

■ Student worksheet
– Why do hydrogen and fluorine react?

Some students may require additional paper to complete a full answer.

Feedback for students

A suitable answer would normally make reference to the free energy change of the reaction (as well as perhaps the activation energy not being prohibitive), and is likely to refer to the bond enthalpies (or strengths) in reactant and product, and the energy changes involved in breaking and forming bonds. A mechanistic answer may refer to the electrical interactions between the cores and electrons in the reactant species. A few students may make reference to the reconfiguration of the charges – *eg* that after reaction the electrons are on the whole more tightly bound in the molecules.

Why do hydrogen and fluorine react?

Hydrogen reacts with fluorine to give hydrogen fluoride. The equation for this reaction is:

$$H_2(g) + F_2(g) \rightarrow 2HF(g)$$

The word equation is:

hydrogen + fluorine → hydrogen fluoride

Look at the following diagrams:

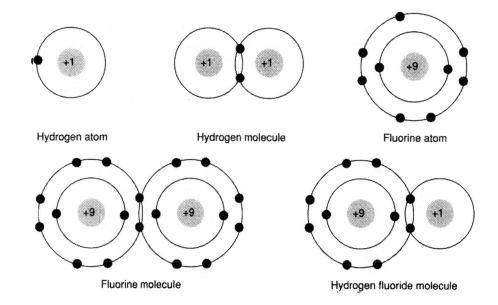

Hydrogen atom   Hydrogen molecule   Fluorine atom

Fluorine molecule   Hydrogen fluoride molecule

In your own words, explain why you think hydrogen reacts with fluorine:

_____

_____

_____

_____

_____

_____

_____

_____

_____

RS•C

RS•C

An analogy for the atom

Target level

This probe is designed for students on post-16 courses, or for students in the 14–16 age range who have already studied the topic of atomic structure.

Topics

The structure of the atom, intra-atomic forces, the 'the atom is a tiny solar system' comparison.

Rationale

Teachers and textbooks often use analogies to introduce unfamiliar ideas. One analogy that is commonly used is that 'the atom is like a tiny solar system'. However, without help, many students have difficulties recognising which aspects of an analogy they are meant to attend to. Students may also be less familiar with the analogue (*eg* the solar system) than is assumed.

This probe will elicit students' ideas about the forces acting in an atom, and in a solar system. It will also provide students with the opportunity to demonstrate their appreciation of the ways in which atoms and solar systems are similar, and the ways they differ. This is to some extent an open-ended task, as there is room for students to use their imagination and creativity to suggest comparisons.

These ideas are discussed in Chapter 7 of the Teachers' notes.

During piloting it was found that some 14–16 year old students find this type of activity very difficult. Some found it difficult to relate what they saw as their physics knowledge to their chemistry knowledge. Some students could not see the point of the exercise - 'they prefer to think of the two things as quite different'. As teachers commonly use analogies and metaphors to help students understand abstract ideas, this exercise could be useful for initiating a discussion of how we use such comparisons in teaching and learning science.

Teachers found that the exercise was fascinating. They were 'amazed' how difficult the exercise was for students, and were surprised by some of the misunderstandings revealed (with post-16 students as well as 14–16 year olds). It was also found that even able students were generally only able to suggest a small number of points when comparing the two systems.

Instructions

Comparing the atom with the solar system is designed to be distributed once students have tackled the questions on **The atom and the solar system** worksheet.

Resources

- Student worksheets
  - The atom and the solar system
  - Comparing the atom with the solar system

Feedback for students

An answer sheet for teachers is provided.

RS•C

An analogy for the atom – answers

The atom and the solar system

1. Electrical

2. The force attracting electron 3 is weaker – as it is a greater distance from the nucleus.

3. The force attracting the nucleus to the electron is the same size as the force attracting the electron to the nucleus – the forces between two bodies always act with the same magnitude (size) on both ('action' = 'reaction').

4. Yes (electrical) – they repel as they both have negative charges.

5. Gravitational

6. The force attracting planet C is weaker – as it is a greater distance from the sun.

7. The force attracting the sun to the planet is the same size as the force attracting the planet to the sun – the forces between two bodies always act with the same magnitude (size) on both ('action' = 'reaction').

8. Yes (gravitational) – they attract as they both have mass.

Comparing the atom with the solar system

Note: that as this is an open-ended activity, other valid ideas should be welcomed. Post-16 students will normally be expected to offer more sophisticated suggestions than younger students (indicated by *). Suggested answers might include:

Similarities
Central body; 'orbiting' * bodies (3 in the examples given); most of mass of system at centre; orbiting bodies attract central body; orbiting bodies attracted by central body; forces act between orbiting bodies; nuclei and sun may have a 'shell' type structure ...

Differences
Size!; nature of attraction (electrical/gravitational); planets do not share 'orbits' * ; electrons repel each other and planets attract each other; many atoms effectively identical – each solar system unique; solar systems evolve whereas atomic transitions are abrupt; electrons may effectively shield part of nuclear charge; all electrons are identical whereas planets are each different (in mass, size, composition); atoms seldom found in a free state, stable solar systems tend to be discrete; nucleus has 'grain' structure (discrete nucleons); solar system mostly in one plane (atoms approximate spherical symmetry); planets may have their own satellites.

* In post-16 courses the idea of electrons 'orbiting' may well have been replaced by more sophisticated ideas about orbitals, electron 'waves' and electron density.

The atom and the solar system

The diagram on the right shows a simple model of an atom.

N is the nucleus, and there are three electrons, labelled 1, 2 and 3.

The electrons are attracted to the nucleus.

Below are some questions about the model of the atom shown in the diagram.

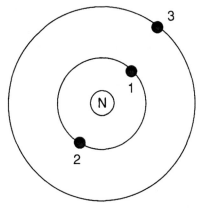

1. What type of force attracts the electrons towards the nucleus? _____

2. Is electron 3 attracted to the nucleus by a stronger force, a weaker force, or the same size force as electron 1?

    _____

    Why do you think this? _____

    _____

    _____

3. Which statement do you think is correct (✔) ?

    ❑   The force attracting the nucleus to electron 2 is larger than the force attracting electron 2 to the nucleus.

    ❑   The force attracting the nucleus to electron 2 is the same size as the force attracting electron 2 to the nucleus.

    ❑   The force attracting the nucleus to electron 2 is smaller than the force attracting electron 2 to the nucleus.

    ❑   There is no force acting on the nucleus attracting it to electron 2.

    Why do you think this? _____

    _____

    _____

4. Is there any force between electron 1 and electron 3? _____

    Why do you think this? _____

    _____

    _____

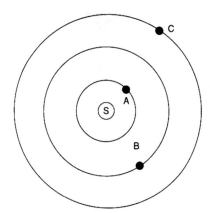

The diagram on the left shows a simple model of a solar system.

S is the sun, and there are three planets, labelled A, B and C.

The planets are attracted to the sun.

Below are some questions about the solar system shown in the diagram.

5. What type of force attracts the planets towards the sun? _____

_____

6. Is planet C attracted to the sun by a stronger force, a weaker force, or the same size force as planet A?

_____

Why do you think this? _____

_____

_____

7. Which statement do you think is correct (✔) ?

❑   The force attracting the sun to planet B is larger than the force attracting planet B to the sun.

❑   The force attracting the sun to planet B is the same size as the force attracting planet B to the sun.

❑   The force attracting the sun to planet B is smaller than the force attracting planet B to the sun.

❑   There is no force acting on the sun attracting it to planet B.

Why do you think this? _____

_____

_____

8. Is there any force between planet A and planet C?

_____

Why do you think this? _____

_____

_____

Comparing the atom with the solar system

Look at the models shown in the diagrams, and try to think of ways in which the atom and the solar system are similar, and ways in which they are different:

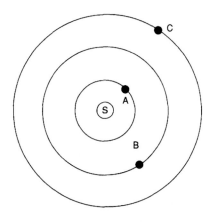
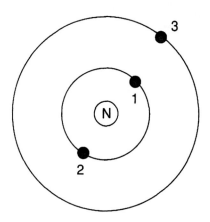

List the similarities and differences you can think of.

In which ways are they similar?

_____

_____

_____

_____

_____

_____

_____

_____

_____

In which ways are they different?

_____

_____

_____

_____

_____

_____

_____

_____

_____

RS•C

This page has been intentionally left blank.

RS•C

Ionisation energy

Target level

This is a diagnostic probe designed for post-16 students studying chemistry.

Topics

Ionisation energy, structure of the atom, intra-atomic forces.

Rationale

This probe is designed to elicit common alternative conceptions about the nature of the interactions between an atomic nucleus and electrons. These incorrect ideas tend to be used by students when explaining patterns in ionisation energy. In particular, students commonly consider the nucleus as giving rise to a set amount of force to be shared among electrons (the 'conservation of force' conception). They may also consider that any species with an octet of electrons in the outer shell must be stable.

These ideas are discussed in Chapter 7 of the Teachers' notes.

During piloting, the probe was described as 'demanding' and 'very challenging' (eg 'it does make the students think more than they usually have to about the subject'), but also as 'helpful' and as a 'useful exercise'. It was suggested that the probe is most valuable as a revision exercise at the end of the topic. Some teachers found that the probe led to useful discussions among students, which were 'thought-provoking' and helped to clarify students' ideas.

Instructions

It is worth pointing out to the students that the statements refer to the diagram. You may wish to also remind students that sometimes several factors may influence an effect, and that there may be several ways of explaining the same effect.

If you do not specifically teach the concept of core charge, then you may wish to suggest that students should omit item 10 or you could modify the worksheet.

A blank answer sheet is provided offering the options 'true', 'false' and 'do not know' for each item. An alternative version of the answer sheet only includes the 'true' and 'false' options as some teachers prefer not to allow a 'do not know' option.

Resources

■ Student worksheets
  – Ionisation energy – true or false?
  – Ionisation energy – answers

Feedback for students

A suggested answer sheet is provided discussing the science behind the statements.

Ionisation energy – answers

Below you will find listed the 20 items you were asked to think about. Following each is a brief comment suggesting whether or not the statement is true, and why.

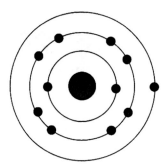

1. Energy is required to remove an electron from the atom.
   True: this is the ionisation energy.

2. After the atom is ionised, it then requires more energy to remove a second electron because the second electron is nearer the nucleus.
   True: When two charged particles interact the force they experience depends on the size of their charges, and their separation. The greater the separation the smaller the force.

3. The atom will spontaneously lose an electron to become stable.
   False: work has to be done to remove an electron – *ie* ionisation energy has to be applied. (This is an endothermic step in the Born-Haber cycle.)

4. Only one electron can be removed from the atom, as it then has a stable electronic configuration.
   False: although the second ionisation energy is considerably larger than the first.

5. The nucleus is not attracted to the electrons.
   False: The nucleus is attracted to an electron with the same magnitude (size) force as the electron is attracted toward the nucleus.

6. Each proton in the nucleus attracts one electron.
   False: All the protons in the nucleus attract all the electrons (and vice versa).

7. After the atom is ionised, it then requires more energy to remove a second electron because the second electron experiences less shielding from the nucleus.
   True: For the outermost electron there are 10 shielding electrons (which repel it, and effectively cancel the attraction due to 10 of the 11 protons in the nucleus), so the core charge is +1, but for the next electron removed there are only 2 shielding electrons so the core charge is +9.

8. The nucleus is attracted towards the outermost electron less than it is attracted towards the other electrons.
   True: as the nucleus is further from the outermost electron. When two charged particles interact the force they experience depends on the size of their charges, and their separation. The same magnitude (size) force acts on both particles, although the forces are in opposite directions. (The force on the electron is directed toward the nucleus, the force on the nucleus acts towards the electron.)

9. After the atom is ionised, it then requires more energy to remove a second electron because the second electron is in a lower energy level.
   True: The outermost electron is a 3s electron, the next to be removed a 2p electron, which is at a lower energy (largely because it is closer to the nucleus, and shielded less and hence attracted more).

10. After the atom is ionised, it then requires more energy to remove a second electron because it experiences a greater core charge than the first.
    True: the first electron experiences a core charge of (11–10=) +1, the second electron experiences a core charge of (11–2=) +9.

11. After the atom is ionised, it then requires more energy to remove a second electron because it would be removed from a positive species.

True: the first electron has to be pulled away from a cation with +1 charge, but the second electron has to be pulled away from a +2 charge, so a greater force has to be overcome.

12. If the outermost electron is removed from the atom it will not return because there will be a stable electronic configuration.

False: unless the electron is attracted somewhere else it will be attracted back to the positive sodium ion.

13. The force on an innermost electron from the nucleus is equal to the force on the nucleus from an innermost electron.

True: the force acting on both has the same magnitude.

14. Electrons do not fall into the nucleus as the force attracting the electrons towards the nucleus is balanced by the force repelling the nucleus from the electrons.

False: as the nucleus is attracted to the electrons, not repelled! The electrons do not fall in to the nucleus as quantum rules only allow the electrons to occupy certain 'positions' *ie* orbitals.

15. The third ionisation energy is greater than the second as there are less electrons in the shell to share the attraction from the nucleus.

False: although the ionisation energy is greater, the attraction from the nucleus is not shared. The actual reasons are (i) there is less repulsion from other electrons counteracting the attraction from the nucleus; and (ii), because of (i), the ionic radius decreases so the third electron (to be removed) is closer to the nucleus.

16. The force pulling the outermost electron towards the nucleus is greater than the force pulling the nucleus towards the outermost electron.

False: Both experience the same magnitude force.

17. After the atom is ionised, it then requires more energy to remove a second electron because once the first electron is removed the remaining electrons receive an extra share of the attraction from the nucleus.

False: more energy is required to remove the second electron as it is closer to the nucleus, experiences a larger core charge, and is being removed from a more positive species. However, the removal of the 3s electron does not increase the attraction to the nucleus experienced by the other electrons.

18. The atom would be more stable if it 'lost' an electron.

False: the separated cation and electron are less stable, which is why energy is required to ionise the atom.

19. The eleven protons in the nucleus give rise to a certain amount of attractive force that is available to be shared between the electrons.

False: the positive charge in the nucleus gives rise a force-field that 'permeates' through space. However there is no force produced unless another charged particle (*eg* an electron) is present in the field. Each electron will interact with the force field due to the nucleus, without 'using it up' in any way. (The electrons will, of course, repel each other - which may counteract the effect of the attraction towards the nucleus to some extent.)

20. The atom would become stable if it either lost one electron or gained seven electrons.

False: although removal of an electron leaves a stable electronic configuration (2.8, isoelectronic with neon), it requires energy as the isolated (negative) electron and the (positive) cation formed would be attracted back together. Gaining seven electrons would give an octet electronic structure (2.8.8, isoelectronic with argon) – however a seven minus ion (Na^{7-}) would not be stable as the repulsions would far outbalance the attractions. There would be eight electrons in the third shell, all repelling each other, and only attracted to a core charge of +1.

Ionisation energy – true or false?

The statements below refer to this diagram of the electronic structure of an atom.

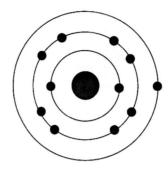

Please read each statement carefully, and decide whether it is correct or not.

1. Energy is required to remove an electron from the atom.

2. After the atom is ionised, it then requires more energy to remove a second electron because the second electron is nearer the nucleus.

3. The atom will spontaneously lose an electron to become stable.

4. Only one electron can be removed from the atom, as it then has a stable electronic configuration.

5. The nucleus is not attracted to the electrons.

6. Each proton in the nucleus attracts one electron.

7. After the atom is ionised, it then requires more energy to remove a second electron because the second electron experiences less shielding from the nucleus.

8. The nucleus is attracted towards the outermost electron less than it is attracted towards the other electrons.

9. After the atom is ionised, it then requires more energy to remove a second electron because the second electron is in a lower energy level.

10. After the atom is ionised, it then requires more energy to remove a second electron because it experiences a greater core charge than the first.

11. After the atom is ionised, it then requires more energy to remove a second electron because it would be removed from a positive species.

12. If the outermost electron is removed from the atom it will not return because there will be a stable electronic configuration.

13. The force on an innermost electron from the nucleus is equal to the force on the nucleus from an innermost electron.

14. Electrons do not fall into the nucleus as the force attracting the electrons towards the nucleus is balanced by the force repelling the nucleus from the electrons.

15. The third ionisation energy is greater than the second as there are less electrons in the shell to share the attraction from the nucleus.

16. The force pulling the outermost electron towards the nucleus is greater than the force pulling the nucleus towards the outermost electron.

17. After the atom is ionised, it then requires more energy to remove a second electron because once the first electron is removed the remaining electrons receive an extra share of the attraction from the nucleus.

18. The atom would be more stable if it 'lost' an electron.

19. The eleven protons in the nucleus give rise to a certain amount of attractive force that is available to be shared between the electrons.

20. The atom would become stable if it either lost one electron or gained seven electrons.

True or false? – response sheet

1.	True	False	1.
2.	True	False	2.
3.	True	False	3.
4.	True	False	4.
5.	True	False	5.
6.	True	False	6.
7.	True	False	7.
8.	True	False	8.
9.	True	False	9.
10.	True	False	10.
11.	True	False	11.
12.	True	False	12.
13.	True	False	13.
14.	True	False	14.
15.	True	False	15.
16.	True	False	16.
17.	True	False	17.
18.	True	False	18.
19.	True	False	19.
20.	True	False	20.

True or false? – response sheet

1.	True	Do not know	False	1.
2.	True	Do not know	False	2.
3.	True	Do not know	False	3.
4.	True	Do not know	False	4.
5.	True	Do not know	False	5.
6.	True	Do not know	False	6.
7.	True	Do not know	False	7.
8.	True	Do not know	False	8.
9.	True	Do not know	False	9.
10.	True	Do not know	False	10.
11.	True	Do not know	False	11.
12.	True	Do not know	False	12.
13.	True	Do not know	False	13.
14.	True	Do not know	False	14.
15.	True	Do not know	False	15.
16.	True	Do not know	False	16.
17.	True	Do not know	False	17.
18.	True	Do not know	False	18.
19.	True	Do not know	False	19.
20.	True	Do not know	False	20.

RS•C

Chemical stability

Target level

This probe may be used at post-16 level, or with students in the 14–16 age range who have studied atomic structure.

Topics

Relative stability of atoms and ions.

Rationale

Research has shown that students entering post-16 chemistry courses may have acquired the notion that species with octet configurations or full outer shells are always more stable than species with other configurations. These judgements are sometimes made regardless of the context, and any other factors that might be relevant. These ideas are discussed in Chapter 6 of the Teachers' notes.

This set of probes asks students to compare the stability of various triads of related species. There are seven probes in the set. The first four compare an atom with two of its ions, and are suitable for able students in the 14–16 year age range as well as for more advanced students. The other three probes introduce more subtle points, and are suitable for students on post-16 courses.

When these materials were piloted teachers found the probes useful and the student responses interesting. It was found that for many students 'the complete shell of electrons was regarded as automatically conferring stability'. One teacher noted that 'it is clear that the complete shell of electrons dominates their thinking. We have spent some time looking at stability in terms of energy changes and forces between charged particles, so it shows me how easily people revert to simple and familiar ideas'.

Instructions

The probes may be used in a number of ways. It is suggested that it is most useful to distribute copies of the various probes in such a way that each student has a different task to their neighbours, and then – after allowing time to complete the sheet - moving to a discussion of the responses (either in groups or in the whole class).

Each student in a 14–16 year old group will require a copy of one of the worksheets:

■ **Chemical stability (1)** – Na^+ / Na / Na^{7-}

■ **Chemical stability (2)** – Cl^{7+} / Cl / Cl^-

■ **Chemical stability (3)** – C^{4+} / C / C^{4-}

■ **Chemical stability (4)** – Be^{2+} / Be / Be^{6-}

Each student in a post-16 group will require a copy of one of the worksheets above, or one of

■ **Chemical stability (5)** – Cl (1.8.8) / Cl (2.7.8) / Cl (2.8.7)

■ **Chemical stability (6)** – Cl / Cl^- / Cl^{11-}

■ **Chemical stability (7)** – O / O^- / O^{2-}

RS•C

Resources

■ Student worksheets
– Chemical stability (1–7)

Feedback for students

A sheet of answers and feedback points for discussion is provided for teachers.

RS•C

RS•C

Chemical stability – answers

Teachers' feedback sheet for class discussion

Note: species with 'octet' of 'full shell' structures tend to predominate within stable chemical systems (molecules, ionic lattices, metallic lattices etc). Students may commonly extrapolate to believe that any species with such a structure (eg Na^{7-}) is always more stable than a species without such a structure. However, when the chemical species are considered in isolation, it is often the case that neutral atoms are more stable than the related ions. This must not be seen as an absolute rule though, as first electron affinities may be exothermic.

Chemical stability (1) – Na^+ / Na / Na^{7-}

1. The neutral atom is more stable, in the sense that the ion will spontaneously attract an electron, but the atom will not spontaneously emit an electron. Students may reasonably argue that they were thinking of real chemical contexts, where the cation is commonly found as part of real chemical substances (sodium metal, sodium chloride etc), unlike the metal. (The separate probe **Stability and reactivity** may be useful to follow up any ambiguity in terms of the context of the question).

2. Clearly the Na^{7-} ion would be highly unstable, but some students will suggest it is more stable than the neutral atom due to its octet structure (full [sic] outer shell). However, in fact it is 10 electrons short of a full outer shell.

3. Clearly the Na^{7-} ion would be highly unstable, but some students will suggest it is equally as stable as the cation due to its octet structure (full [sic] outer shell).

Chemical stability (2) – Cl^{7+} / Cl / Cl^-

4. A great deal of energy is required to ionise a chlorine atom to give a 7+ cation, but some students may feel the cation is more stable due to its octet structure.

5. Chlorine has an exothermic electron affinity, so the chloride ion may be considered more stable than the atom, despite the atom's neutrality.

6. A great deal of energy is required to ionise a chlorine atom to give a 7+ cation, but some students may feel the two ions are equally stable due to their octet structures. (It is possible some may suggest that the cation is more stable, as it has a full outer shell - although many students are likely to assign full shells to both of these ions.)

Chemical stability (3) – C^{4+} / C / C^{4-}

7. A great deal of energy is required to ionise a carbon atom to give a 4+ cation, but some students may feel the cation is more stable due to its full shell structure.

8. The C^{4-} ion would be unstable, but some students will suggest it is more stable than the neutral atom due to its octet structure (full [sic] outer shell).

9. Both of the ions would be highly labile. Some students may feel both are stable because they have full shells, whilst others may suggest that the anion is more stable as it has an octet, or because it has more full shells.

RS•C

Chemical stability (4) – Be^{2+} / Be / Be^{6-}

10. Energy is required to ionise the atom to form the cation, so the neutral atom should be considered more stable. (However, see the comments on Question 1 about the context in which students might consider the question.)

11. The highly charged anion is clearly unstable compared the neutral atom, although some students may feel that the octet structure on the anion makes it more stable.

12. The highly charged metal anion is clearly unstable compared the moderately charged metal cation. Some students may feel both are stable because they have full shells, whilst others may suggest that the anion is more stable as it has an octet, or because it has more full shells.

Chemical Stability (5) – Cl (1.8.8) / Cl (2.7.8) / Cl (2.8.7)

13. Neither of these configurations are stable, and both would be expected to undergo spontaneous transition to 2.8.7. The 1.8.8 structure will emit a greater energy quanta as the electron drops from the n=3 to the n=1 level (cf the quantum jump from n=3 to n=2 in the 2.7.8 species). Some students may feel that both are stable as they have octets in the outer shells, and 1.8.8 may be considered more stable as it has two octets.

14. 2.8.7 is the ground state of the chlorine atom and is more stable. The excited state (2.7.8) will decay to the ground state. Some students may feel that the excited state is more stable as it has an octet in the outer shell.

15. 2.8.7 is the ground state of the chlorine atom and is more stable. The excited state (1.8.8) will decay to the ground state. Some students may feel that the excited state is more stable as it has an octet in the outer shell.

Chemical stability (6) – Cl / Cl^- / Cl^{11-}

16. Chlorine has an exothermic electron affinity, so the chloride ion may be considered more stable than the atom, despite the atom's neutrality.

17. The common Cl^- anion is clearly more stable than the highly charged Cl^{11-} ion. Cl^- has an outer octet of electrons (but not a full shell), and Cl^{11-} has a full outer shell (but more than an octet).

18. The neutral atom is clearly more stable than the highly charged Cl^{11-} ion. However, Cl^{11-} has a full outer shell and may be considered more stable by some students.

Chemical stability (7) – O / O^- / O^{2-}

19. Oxygen has an exothermic electron affinity, so the O^- ion may be considered more stable than the atom, despite the atom's neutrality.

20. Although the oxide is O^{2-} and is common, the O^- anion will actually repel away another electron, so is more stable than O^{2-}. (See the comments about chemical context in Q1). The electron affinity for the oxygen atom is -142 kJ mol^{-1}, but the electron 'affinity' [sic] of the O^- ion is endothermic: +844 kJ mol^{-1}.

21. The O^{2-} ion, although common is, in isolation, less stable than the neutral atom. (The sum of the two electron affinity values is endothermic – see Question 20.)

Chemical stability (1)

The diagrams below represent three chemical species:-

Na$^+$

Na

Na^{7-}

Sodium ion with
electronic configuration
of 2.8

Sodium atom with
electronic configuration
of 2.8.1

Sodium ion with
electronic configuration
of 2.8.8

1. Tick ✔ one of the four statements:

❑   Na$^+$ is more stable than Na

❑   Na$^+$ and Na are equally stable

❑   Na$^+$ is less stable than Na

❑   I do not know

Why did you think this was the answer?

_____

_____

_____

_____

2. Tick ✔ one of the four statements:

❑   Na is more stable than Na^{7-}

❑   Na and Na^{7-} are equally stable

❑   Na is less stable than Na^{7-}

❑   I do not know

Why did you think this was the answer?

_____

_____

_____

_____

3. Tick ✔ one of the four statements:

❑   Na^{7-} is more stable than Na$^+$

❑   Na^{7-} and Na$^+$ are equally stable

❑   Na^{7-} is less stable than Na$^+$

❑   I do not know

Why did you think this was the answer?

_____

_____

_____

_____

Chemical stability (2)

The diagrams below represent three chemical species:

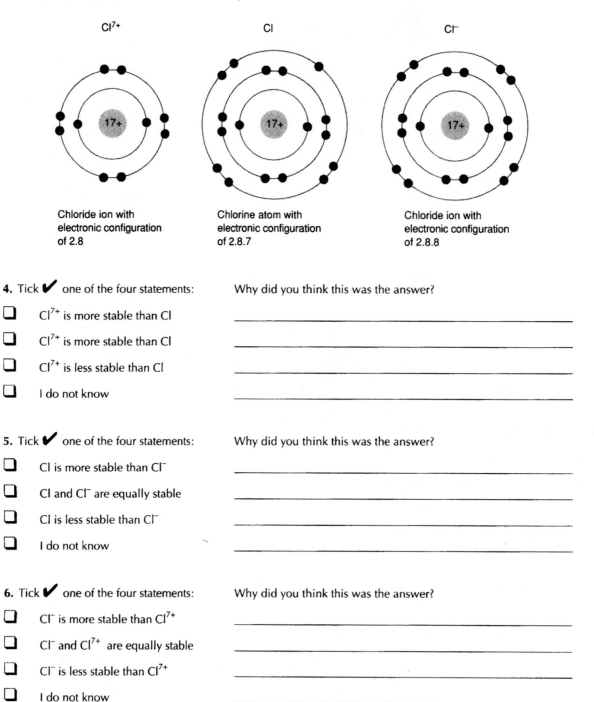

Cl^{7+}

Cl

Cl$^-$

Chloride ion with electronic configuration of 2.8

Chlorine atom with electronic configuration of 2.8.7

Chloride ion with electronic configuration of 2.8.8

4. Tick ✔ one of the four statements:

❑   Cl^{7+} is more stable than Cl

❑   Cl^{7+} is more stable than Cl

❑   Cl^{7+} is less stable than Cl

❑   I do not know

Why did you think this was the answer?

_____

_____

_____

_____

5. Tick ✔ one of the four statements:

❑   Cl is more stable than Cl$^-$

❑   Cl and Cl$^-$ are equally stable

❑   Cl is less stable than Cl$^-$

❑   I do not know

Why did you think this was the answer?

_____

_____

_____

_____

6. Tick ✔ one of the four statements:

❑   Cl$^-$ is more stable than Cl^{7+}

❑   Cl$^-$ and Cl^{7+} are equally stable

❑   Cl$^-$ is less stable than Cl^{7+}

❑   I do not know

Why did you think this was the answer?

_____

_____

_____

_____

RS•C

Chemical stability (3)

The diagrams below represent three chemical species:-

C⁴⁺   C   C⁴⁻

Carbon ion with electronic
configuration of 2

Carbon atom with electronic
configuration of 2.4

Carbon ion with electronic
configuration of 2.8

7. Tick ✔ one of the four statements:

❑ C^{4+} is more stable than C

❑ C^{4+} and C are equally stable

❑ C^{4+} is less stable than C

❑   I do not know

Why did you think this was the answer?

_____

_____

_____

_____

8. Tick ✔ one of the four statements:

❑   C is more stable than C^{4-}

❑   C and C^{4-} are equally stable

❑   C is less stable than C^{4-}

❑   I do not know

Why did you think this was the answer?

_____

_____

_____

_____

9. Tick ✔ one of the four statements:

❑ C^{4-} is more stable than C^{4+}

❑ C^{4-} and C^{4+} are equally stable

❑ C^{4-} is less stable than C^{4+}

❑   I do not know

Why did you think this was the answer?

_____

_____

_____

_____

Chemical stability (4)

The diagrams below represent three chemical species:-

Be^{2+} Be Be^{6-}

Beryllium ion with electronic configuration of 2

Beryllium atom with electronic configuration of 2.2

Beryllium ion with electronic configuration of 2.8

10. Tick ✔ one of the four statements:

☐ Be^{2+} is more stable than Be

☐ Be^{2+} and Be are equally stable

☐ Be^{2+} is less stable than Be

☐   I do not know

Why did you think this was the answer?

_____

_____

_____

_____

11. Tick ✔ one of the four statements:

☐   Be is more stable than Be^{6-}

☐   Be and Be^{6-} are equally stable

☐   Be is less stable than Be^{6-}

☐   I do not know

Why did you think this was the answer?

_____

_____

_____

_____

12. Tick ✔ one of the four statements:

☐ Be^{6-} is more stable than Be^{2+}

☐ Be^{6-} and Be^{2+} are equally stable

☐ Be^{6-} is less stable than Be^{2+}

☐   I do not know

Why did you think this was the answer?

_____

_____

_____

_____

Chemical stability (5)

The diagrams below represent three chemical species:-

1.8.8

2.7.8

2.8.7

Chlorine atom with
electronic configuration
of 1.8.8

Chlorine atom with
electronic configuration
of 2.7.8

Chlorine atom with
electronic configuration
of 2.8.7

13. Tick ✔ one of the four statements:

❑ 1.8.8 is more stable than 2.7.8

❑ 1.8.8 and 2.7.8 are equally stable

❑ 1.8.8 is less stable than 2.7.8

❑ I do not know

Why did you think this was the answer?

_____

_____

_____

_____

14. Tick ✔ one of the four statements:

❑ 2.7.8 is more stable than 2.8.7

❑ 2.7.8 and 2.8.7 are equally stable

❑ 2.7.8 is less stable than 2.8.7

❑ I do not know

Why did you think this was the answer?

_____

_____

_____

_____

15. Tick ✔ one of the four statements:

❑ 2.8.7 is more stable than 1.8.8

❑ 2.8.7 and 1.8.8 are equally stable

❑ 2.8.7 is less stable than 1.8.8

❑ I do not know

Why did you think this was the answer?

_____

_____

_____

_____

Chemical stability (6)

The diagrams below represent three chemical species:

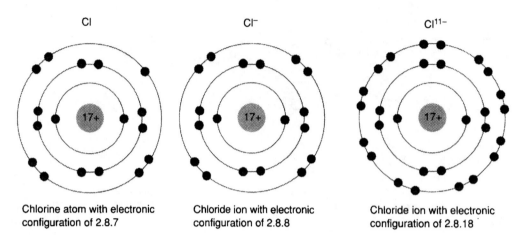

Chlorine atom with electronic configuration of 2.8.7

Chloride ion with electronic configuration of 2.8.8

Chloride ion with electronic configuration of 2.8.18

16. Tick ✔ one of the four statements:

❑   Cl is more stable than Cl⁻

❑   Cl and Cl⁻ are equally stable

❑   Cl is less stable than Cl⁻

❑   I do not know

Why did you think this was the answer?

_____

_____

_____

_____

17. Tick ✔ one of the four statements:

❑   Cl⁻ is more stable than Cl¹¹⁻

❑   Cl⁻ and Cl¹¹⁻ are equally stable

❑   Cl⁻ is less stable than Cl¹¹⁻

❑   I do not know

Why did you think this was the answer?

_____

_____

_____

_____

18. Tick ✔ one of the four statements:

❑   Cl¹¹⁻ is more stable than Cl

❑   Cl¹¹⁻ and Cl are equally stable

❑   Cl¹¹⁻ is less stable than Cl

❑   I do not know

Why did you think this was the answer?

_____

_____

_____

_____

Chemical stability (7)

The diagrams below represent three chemical species:

O   O⁻   O²⁻

Oxygen atom with electronic
configuration of 2.6

Oxygen ion with electronic
configuration of 2.7

Oxide ion with electronic
configuration of 2.8

19. Tick ✔ one of the four statements:

❑   O is more stable than O⁻

❑   O and O⁻ are equally stable

❑   O is less stable than O⁻

❑   I do not know

Why did you think this was the answer?

_____

_____

_____

_____

20. Tick ✔ one of the four statements:

❑   O⁻ is more stable than O²⁻

❑   O⁻ and O²⁻ are equally stable

❑   O⁻ is less stable than O²⁻

❑   I do not know

Why did you think this was the answer?

_____

_____

_____

_____

21. Tick ✔ one of the four statements:

❑   O²⁻ is more stable than O

❑   O²⁻ and O are equally stable

❑   O²⁻ is less stable than O

❑   I do not know

Why did you think this was the answer?

_____

_____

_____

_____

RS•C

This page has been intentionally left blank.

RS•C

Stability and reactivity

Target level

This is a probe which may be used in post-16 courses, or in the later stage of courses for 14–16 year olds.

Topics

This probe is designed to elicit students' ideas about chemical stability and reactivity.

Rationale

This probe is inspired by research which suggests that students often believe that atoms will spontaneously form species with octet structures regardless of the chemical context. It is also considered that students do not logically relate the reactivity of a substance to the stability of the chemical species it is comprised of. This may be, in part, because of the difficulty students have in relating the molar phenomena of chemical reactions with the theoretical model of particle interactions. The items on this probe are intended to act as a starting point for discussing these ideas in the classroom.

Difficulties in applying molecular level models, and learners' ideas about reactions and stability are discussed in Chapter 6 of the Teachers' notes.

Question 2 originally suggested that 'the sodium atom will emit an electron...', but the term 'emit' was considered unfamiliar to many students. The term 'eject' is considered more familiar (*eg* from operating stereo equipment). (Another alternative - that 'the atom would 'lose' an electron' does not necessarily imply a spontaneous process.)

During piloting, teachers noted that the probe 'gave the pupils more opportunity to air their ideas'. One teacher described it as 'good, short, sharp - did not disrupt lesson, yet caused pupils to question their understanding.' One teacher noted that all the students in a post-16 class answered 'in terms of obtaining a full electron shell by losing one electron to obtain a stable octet'. Another noted 'a constant contradiction throughout [the answers from a class of 15–16 year olds] - the atom being most stable yet most willing to donate an electron', and concluded that it was 'possible that students misunderstand the concept of stability'.

Instructions

It is worth emphasising that the questions relate to the comparison shown in the diagram.

Resources

■ Student worksheet
    – Stability and reactivity

Feedback for students

Answers and some suggested discussion points are provided for teachers.

RS•C

Stability and reactivity – answers

In discussing the responses it is suggested that the teacher should focus on the figure on the sheet, and begin by considering question 2. The electron is negatively charged and the ion positive, so they will be spontaneously attracted to form an atom. The atom will not emit an electron without an input of energy. In the context of the diagram, then, the atom is more stable (question 1), and the ion is more 'reactive' (question 3).

It should be borne in mind, however, that this exercise is intended as a probe, and not a test. Some students are likely to think in terms of more familiar chemical contexts, *eg* comparing a sodium ion in sodium chloride with elemental sodium (which students may consider to contain atoms), where the metal is part of a more reactive system in the presence of water, or air etc. An important teaching point is that students need to consider the overall chemical context when judging 'stability' or 'reactivity'.

1. The sodium atom is more stable than the sodium ion.

2. The sodium ion and electron will combine to become an atom.

3. The sodium ion is more reactive than the sodium atom.

Stability and reactivity

The three questions all refer to the diagram on the right.

Please tick one box for each of the three questions, and explain your reasons.

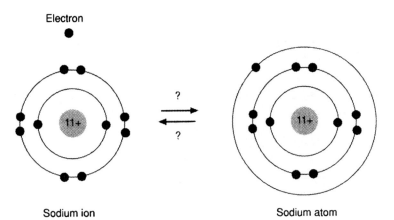

Electron

Sodium ion

Sodium atom

Question 1

✔

Reason

☐ The sodium atom is more stable than the sodium ion. _____

☐ The sodium ion is more stable than the sodium atom. _____

☐ The sodium ion and sodium atom are equally stable. _____

☐ I do not know which statement is correct. _____

Question 2

✔

Reason

☐ The sodium atom will eject (give out) an electron to become an ion.

_____

☐ The sodium ion and electron will combine to become an atom.

_____

☐ Neither of the changes suggested above will occur. _____

☐ I do not know which statement is correct. _____

Question 3

✔

Reason

☐ The sodium atom is more reactive than the sodium ion. _____

☐ The sodium ion is more reactive than the sodium atom. _____

☐ The sodium ion and sodium atom are equally reactive. _____

☐ I do not know which statement is correct. _____

RS•C

This page has been intentionally left blank.

RS•C

RS•C

Interactions

Target level

This probe is intended for students studying chemistry on post-16 courses.

Topics

Chemical bonding (including: ionic, covalent, metallic, polar, hydrogen, dipole-dipole, van der Waals, solvation, dative, double).

Rationale

Research suggests that students commonly focus on covalent and ionic bonding, and often fail to spot, or may down-play, the importance of other types of bonding. Students may have idiosyncratic notions of the distinction between the terms 'force', 'attraction', 'bonding' and 'chemical bond'. These ideas are discussed in Chapter 8 of the Teachers' notes. This probe provides a way of exploring students' ideas about different bond types in some detail. (The probe **Spot the bonding** will provide a quicker means of auditing which bond types students can identify.)

A variety of types of diagram are used in this probe, as it is important for students to be able to interpret and use various ways of representing chemical species (see Chapter 6 of the Teachers' notes).

During piloting, this worksheet was described as a 'good probe of misconceptions' which 'hit the spot'. Students reported that the exercise made them think, and made them aware of ideas that they needed to revise.

Resources

■ Student worksheets
   – Interactions

Feedback for students

A generous space allowance is provided, and it should be emphasized that students do not need to fill up all the space available.

A suggested answer sheet is provided for teachers.

RS•C

Interactions – answers

Classifying interactions as attraction, force, bonding or chemical bond

In question 1: the interaction between the nucleus and the electron is an attraction, based on electrical forces, but is not usually referred to as bonding (although the term binding may sometimes be used), and is not classified as a chemical bond.

In questions 2–10: the interactions shown are examples of bonding and can be classed as chemical bonds, as well as being attractions based upon electrical forces.

(Teachers will have their own expectations about the level of detail appropriate in the responses from their groups.)

1. No specific name – intra-atomic forces or just electrical forces/electrostatic
    Description – electrical attraction between positive nucleus and negative electron.

2. Covalent bond
    Description – a negative pair of electrons between the two positive nuclei binds them together. Some students may discuss the formation of a (molecular) bonding orbital from the overlap of atomic orbitals.

3. **a) and b)** Ionic bond
    Description – electrical attraction between each ion and the surrounding counter ions – *ie* positive cations attracted to and by surrounding negative anions – and vice versa.

4. Hydrogen bonding (and van der Waals forces and dipole-dipole forces)
    Description – due to the difference in electronegativity between oxygen and hydrogen the bonds in water are polar ($H^{\delta+}$ and $O^{\delta-}$). Hydrogen bonding is formed when the (δ^+) hydrogen centre of one molecule is attracted to and attracts the (δ^-) oxygen centre of another. A lone pair (*ie* non-bonding pair) of electrons on one molecule attracts and is attracted by a poorly shielded proton on another molecule.

5. Solvent-solute interactions/solvation forces/hydration forces
    Description – the negative poles of water molecules attract and are attracted to, the positive cations (and the positive poles of water molecules attract, and are attracted to, the negative anion).

6. Covalent bond
    Description – a negative pair of electrons between the two positive cores binds them together.
    Some students may discuss the formation of a (molecular) bonding orbital from the overlap of atomic orbitals.

7. Van der Waals forces
    Description – the synchronisation of transient fluctuating dipoles leads to induced dipole - induced dipole forces between molecules.

8. Metallic
    Description – the delocalised electrons attract, and are attracted to, the positive atomic cores.
    (Some students may refer to the overlap of atomic orbitals to form extensive molecular orbitals – *ie* the conduction band. Other students may be aware that bonding in transition metals can be considered to have covalent character in addition to its metallic nature.)

9. Dative /co-ordinate bonding
    Description – a lone pair of electrons on the chlorine centre from one molecule attracts, and is attracted by, the poorly shielded positively charged aluminium core of the other molecule – and vice versa.

10. Covalent bond
    Description – a negative pair of electrons between the two adjacent positive carbon cores binds them together.
    Some students may discuss the formation of (molecular) bonding orbitals from the overlap of atomic orbitals.

Notes

a) At equilibrium the forces (attractions and repulsions) in molecules etc are in balance.

b) Question 3 is reproduced in two versions (3a and 3b) to allow students to demonstrate that they understand that the bonding involves the interactions between each ion each of its surrounding counter ions. Some students may class 3a as a bond, but suggest 3b (involving the same cation) must be just a force. (See Chapter 6 of the Teachers' notes.)

Interactions

This exercise is about the interactions between particles in different chemical systems (such as molecules, atoms, lattices). It has been found that different students have distinct ideas about how to label and describe the interactions that are found in chemical systems.

On each page you will find a diagram representing a chemical system (such as a single molecule, or part of a solid).

You are asked to identify and describe any interactions that are present between different parts of the chemical system in the diagram.

In each case you are asked whether you think that the interaction should be classed as an attraction, a force, bonding and/or a chemical bond. Please tick the 'yes', 'no', or 'unsure' (if you do not know) box for each of these labels. You do not have to select one and only one 'yes' answer: you may select 'no' for all four options, or 'yes' for all four, or any combination of 'yes' and 'no' responses.

You are also asked to label the type of interaction, if you think it has a special name, and to describe the interaction as best you can in your own words. (You do not need to fill up all the lines.)

1. The diagram on the right represents a single atom of hydrogen.

   Which, if any, of the following labels can be used to identify the interaction between the two parts of the system shown:

	Yes?	No?	Unsure?
Attraction	❑	❑	❑
Force	❑	❑	❑
Bonding	❑	❑	❑
Chemical bond	❑	❑	❑

   (please tick one box in each row)

   Do you think this type of interaction is given a particular name/label? (If so, how would you label this type of interaction?)

   _____

   Describe this interaction in your own words. Give as much detail as you can:

   _____
   _____
   _____
   _____
   _____
   _____
   _____
   _____
   _____
   _____
   _____
   _____
   _____

   (You do not need to fill up all the lines.)

2. The diagram on the right represents a single molecule of hydrogen.
Which, if any, of the following labels can be used to identify the interaction between the two parts of the system shown:

	Yes?	No?	Unsure?
Attraction	❑	❑	❑
Force	❑	❑	❑
Bonding	❑	❑	❑
Chemical bond	❑	❑	❑

(please tick one box in each row)

Do you think this type of interaction is given a particular name/label? (If so, how would you label this type of interaction?)

_____

Describe this interaction in your own words. Give as much detail as you can:

_____

_____

_____

_____

_____

_____

(You do not need to fill up all the lines.)

3a. The diagram on the right represents part of a layer in a sodium chloride lattice.
Which, if any, of the following labels can be used to identify the interaction between the two parts of the system shown:

	Yes?	No?	Unsure?
Attraction	❑	❑	❑
Force	❑	❑	❑
Bonding	❑	❑	❑
Chemical bond	❑	❑	❑

(please tick one box in each row)

Do you think this type of interaction is given a particular name/label? (If so, how would you label this type of interaction?)

_____

Describe this interaction in your own words. Give as much detail as you can:

_____

_____

_____

_____

_____

_____

(You do not need to fill up all the lines.)

3b. The diagram on the right represents the same part of a layer in a sodium chloride lattice as the previous question. Which, if any, of the following labels can be used to identify the interaction between the two parts of the system shown:

	Yes?	No?	Unsure?
Attraction	❑	❑	❑
Force	❑	❑	❑
Bonding	❑	❑	❑
Chemical bond	❑	❑	❑

(please tick one box in each row)

Do you think this type of interaction is given a particular name/label? (If so, how would you label this type of interaction?)

_____

Describe this interaction in your own words. Give as much detail as you can:

_____

_____

_____

_____

_____

_____

(You do not need to fill up all the lines.)

4. The diagram on the right represents some water molecules in liquid water. Which, if any, of the following labels can be used to identify the interactions between molecules in the liquid:

	Yes?	No?	Unsure?
Attraction	❑	❑	❑
Force	❑	❑	❑
Bonding	❑	❑	❑
Chemical bond	❑	❑	❑

(please tick one box in each row)

Do you think this type of interaction is given a particular name/label? (If so, how would you label this type of interaction?)

_____

Describe this interaction in your own words. Give as much detail as you can:

_____

_____

_____

_____

_____

_____

(You do not need to fill up all the lines.)

5. The diagram on the right represents part of an aqueous solution of silver nitrate.
Which, if any, of the following labels can be used to identify the interactions between the ions and the molecules in the liquid:

	Yes?	No?	Unsure?
Attraction	❑	❑	❑
Force	❑	❑	❑
Bonding	❑	❑	❑
Chemical bond	❑	❑	❑

(please tick one box in each row)

Do you think this type of interaction is given a particular name/label? (If so, how would you label this type of interaction?)

_____

Describe this interaction in your own words. Give as much detail as you can:

_____

_____

_____

_____

_____

(You do not need to fill up all the lines.)

6. The diagram on the right represents a molecule of fluorine.
Which, if any, of the following labels can be used to identify the interactions which hold the molecule together?

	Yes?	No?	Unsure?
Attraction	❑	❑	❑
Force	❑	❑	❑
Bonding	❑	❑	❑
Chemical bond	❑	❑	❑

(please tick one box in each row)

Do you think this type of interaction is given a particular name/label? (If so, how would you label this type of interaction?)

_____

Describe this interaction in your own words. Give as much detail as you can:

_____

_____

_____

_____

_____

(You do not need to fill up all the lines.)

7. The diagram on the right represents iodine molecules in solid iodine
Which, if any, of the following labels can be used to identify the interactions between the molecules?

	Yes?	No?	Unsure?
Attraction	❏	❏	❏
Force	❏	❏	❏
Bonding	❏	❏	❏
Chemical bond	❏	❏	❏

(please tick one box in each row)

Do you think this type of interaction is given a particular name/label? (If so, how would you label this type of interaction?)

_____

Describe this interaction in your own words. Give as much detail as you can:

_____
_____
_____
_____
_____
_____

(You do not need to fill up all the lines.)

8. The diagram on the right represents the lattice arrangement in copper.
Which, if any, of the following labels can be used to identify the interactions holding the copper together?

	Yes?	No?	Unsure?
Attraction	❏	❏	❏
Force	❏	❏	❏
Bonding	❏	❏	❏
Chemical bond	❏	❏	❏

(please tick one box in each row)

Do you think this type of interaction is given a particular name/label? (If so, how would you label this type of interaction?)

_____

Describe this interaction in your own words. Give as much detail as you can:

_____
_____
_____
_____
_____
_____

(You do not need to fill up all the lines.)

RS•C

9. The diagram on the right represents a dimer of aluminium chloride ($AlCl_3$).
Which, if any, of the following labels can be used to identify the interactions between the two $AlCl_3$ molecules?:

	Yes?	No?	Unsure?
Attraction	❏	❏	❏
Force	❏	❏	❏
Bonding	❏	❏	❏
Chemical bond	❏	❏	❏

(please tick one box in each row)

Do you think this type of interaction is given a particular name/label? (If so, how would you label this type of interaction?)

_____

Describe this interaction in your own words. Give as much detail as you can:

_____

_____

_____

_____

_____

_____

(You do not need to fill up all the lines.)

10. The diagram on the right represents part of the diamond structure of carbon.
Which, if any, of the following labels can be used to identify the interactions holding the structure together?:

	Yes?	No?	Unsure?
Attraction	❏	❏	❏
Force	❏	❏	❏
Bonding	❏	❏	❏
Chemical bond	❏	❏	❏

(please tick one box in each row)

Do you think this type of interaction is given a particular name/label? (If so, how would you label this type of interaction?)

_____

Describe this interaction in your own words. Give as much detail as you can:

_____

_____

_____

_____

_____

_____

(You do not need to fill up all the lines.)

RS•C

RS•C

This page has been intentionally left blank.

RS•C

RS•C

Acid strength

Target level

This is a probe and exercise designed for use in courses with post-16 students.

Topic

Acid solutions - distinguishing between concentration and strength.

Rationale

As in everyday life solutions of differing concentration are described as 'strong' and 'weak', students will often see these terms as being synonymous with 'concentrated' and 'dilute'. The probe is designed to diagnose whether students distinguish strength from concentration. For those that are not clear, the exercise provides practice at discriminating along the two separate dimensions. This is achieved by the use of large and detailed diagrams representing solutions with the four combinations stronger/weaker and more/less concentrated.

These ideas are discussed in Chapter 2 of the Teachers' notes.

During piloting the exercise was described as 'self-explanatory', 'very helpful' and a 'useful teaching instrument'. It was found that the exercise was considered 'useful' by those who were initially unsure of their ideas. A few students may find the exercise too easy, and teachers may wish to use the probe to determine which students will benefit from the exercise. However most students who thought they did understand the ideas found the exercise 'reassuring', and it was considered to provide useful reinforcement of the distinction.

Instructions

This section uses two resources, a probe and an exercise.

Resources

■ Student worksheets
  – Explaining acid strength (probe)
  – Classifying acid solutions (exercise)

Feedback for students

An answer sheet for teachers is provided.

RS•C

Acid strength – answers

Explaining acid strength

1. In a strong acid all (or virtually all) of the molecules dissociate to give hydrated hydrogen ions and anions in solution. In a weak acid, only a proportion of the molecules dissociate to give hydrated hydrogen ions and anions in solution – so there are solvated molecules present as well as solvated ions.

2. A concentrated acid has a relatively large amount of solute dissolved in the solvent.
   A dilute acid has a relatively smaller amount of solute dissolved in the solvent.

3. In a solution of a strong acid there would be solvated ions present, but no (or only a very small proportion of) associated molecules present. In a solution of a weak acid there would be a significant proportion of molecules present in their original form, as well as some ions formed by dissociation of molecules.

Classifying acid solutions

1. Water molecules, hydrated hydrogen ions (*ie* hydroxonium ions), anions (*ie* acid radicals) or H_2O, H_3O^+, A^-.

2. The diagram shows a solution of a strong acid.

3. Water molecules, molecules of the acid, hydrated hydrogen ions (*ie* hydroxonium ions), anions (*ie* acid radicals), or H_2O, HA, H_3O^+, A^-.

4. This diagram shows a solution of a weaker acid than that shown in diagram 1.

5. Water molecules, molecules of the acid, hydrated hydrogen ions (ie hydroxonium ions), anions (*ie* acid radicals) or H_2O, HA, H_3O^+, A^-.

6. This diagram shows a weaker acid than that shown in diagram 1.
   This diagram shows a more concentrated acid than that shown in diagrams 1 and 2.

7. Water molecules, hydrated hydrogen ions (*ie* hydroxonium ions), anions (*ie* acid radicals), or H_2O, H_3O^+, A^-.

8. This diagram shows an acid which is more concentrated than that shown in diagram 1 or 2.
   This diagram shows an acid which is stronger than that shown in diagrams 2 or 3.

9.

	Strong	Weak
Concentrated	4	3
Dilute	1	2

Explaining acid strength

One definition of an acid is that it dissolves in water to give hydrogen ions (H^+).

In fact the hydrogen ion (H^+) will associate with a water molecule to form H_3O^+.

One way to write the equation for an acid 'HA' dissolving in water is:

$$HA + H_2O(l) \rightleftharpoons H_3O^+(aq) + A^-(aq)$$

The A in HA does not stand for a particular element, but for the 'acid radical' part of the molecule. So, for example, in hydrochloric acid 'HA' would be HCl, and 'A$^-$' would be Cl$^-$, whilst in ethanoic acid 'HA' would be CH_3COOH, and 'A$^-$' would be CH_3COO^-.

Acids (and alkalis) can be described as 'strong' or 'weak', and as 'concentrated' or 'dilute'.

1. What is the difference between a strong acid and a weak acid?

2. What is the difference between a concentrated acid and a dilute acid?

3. If you could see the particles (molecules, ions etc) in an acidic solution, how would you decide whether it was a solution of a strong acid or a solution of a weak acid?

Classifying acid solutions

One way to write the equation for an acid 'HA' dissolving in water is:

$$HA + H_2O(l) \rightleftharpoons H_3O^+(aq) + A^-(aq)$$

On the following pages are some diagrams of acidic solutions.

Scientific diagrams are always simplifications designed to highlight some aspects of the system represented. The diagrams in this exercise show simplifications of real solutions.

For example, the concentration of acids varies over many orders of magnitude, and an accurate diagram of a very dilute solution would need to show many thousands of water molecules for each $H^+(aq)$ ion.

Only four types of particle are shown in these diagrams. The following key is used to distinguish between the different particles:

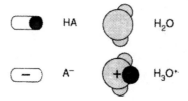

The size (and shape) of acid molecules varies greatly, and they are often much larger than a water molecule.

Look carefully at the four diagrams on the following pages, and see if you can tell what the differences between them are meant to indicate.

RS•C

Diagram 1

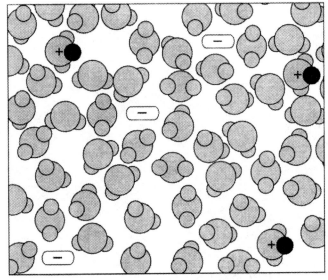

1. What types of particles are shown in the solution represented in this diagram?

   _____

   _____

2. How would you describe this solution?

   _____

   _____

Diagram 2

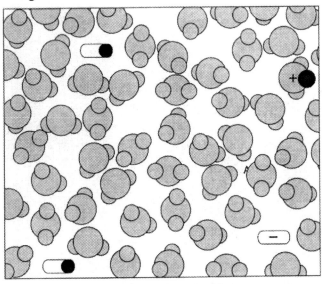

3. What types of particles are shown in the solution represented in this diagram?

   _____

   _____

4. How would you describe this solution (compared to diagram 1)?

   _____

   _____

Diagram 3

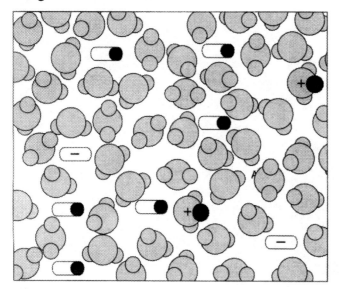

5. What types of particles are shown in the solution represented in this diagram?

_____

_____

6. How would you describe this solution (compared to diagrams 1 and 2)?

_____

_____

Diagram 4

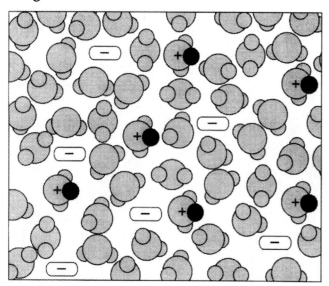

7. What types of particles are shown in the solution represented in this diagram?

_____

_____

8. How would you describe this solution (compared to diagrams 1-3)?

_____

_____

9. The four diagrams you were asked to consider are reproduced in miniature below.

1.   2.

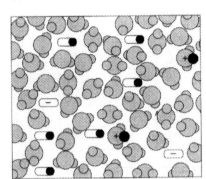

3.   4.

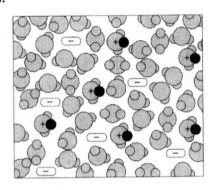

The diagrams are meant to represent a concentrated solution of a strong acid, a dilute solution of a strong acid, a concentrated solution of a weak acid and a dilute solution of a weak acid. Use the table below to show which diagram is meant to represent each of the four solutions – write the number of the appropriate diagram in each box.

	Strong	Weak
Concentrated		
Dilute		

RS•C

This page has been intentionally left blank.

RS•C

RS•C

Reaction mechanisms

Target level

This probe is intended for students on post-16 courses.

Topic

Reaction mechanisms

Rationale

The comprehension of, and use of, reaction mechanisms requires students to not only appreciate the electrical changes which are conjectured to accompany chemical reactions, but also to learn a new formalism where the movement of electrons or electron pairs is represented by full or half arrows ('curly arrows' and 'fish-hooks') which may start and end on either atoms or bonds.

These ideas are discussed in Chapter 9 of the Teachers' notes.

During piloting, it was found that the probe was 'very worthwhile', 'useful for revision' and a 'good way to check understanding of principles'. It was reported that the probe discriminated well between students with a good understanding and weaker candidates. Students generally found it easier to select the correct response than to explain why it was correct (this was considered to demonstrate rote learning without deep understanding). The range of alternative answers was considered to be useful, as it allows valuable discussion of why each wrong answer was not correct.

Instructions

Some teachers may wish to use only one of the question sheets depending upon the examination specification being followed.

It may be worth emphasising to the students that the first step is shown in the centre, and they must select one answer from the surrounding options.

Resources

- Student worksheets
  - Reaction mechanisms – instruction sheet for students
  - Reaction mechanism 1 – question based on an ionic mechanism
  - Reaction mechanism 2 – question based on a free radical mechanism
  - Reaction mechanisms revealed – answers sheet

Feedback for students

An answer sheet, **Reaction mechanisms revealed**, that teachers may wish to issue to students during, or following, discussion of the answers, is provided.

A web site containing tutorials to support the ideas introduced here can be found at **http://www.abdn.ac.uk/curly-arrows/** (accessed October 2001).

Reaction mechanisms

Chemists use reaction mechanisms to show what they think might be happening as molecules interact during chemical reactions.

When drawing reaction mechanisms the chemist usually assumes:

1. that the reaction occurs in several distinct steps;

2. that each step can be represented as the movement of electrons; and

3. that sometimes electrons move as pairs, and sometimes they move individually.

Diagrams showing the steps in reaction mechanisms usually show the molecules and/or ions (shown by + and –) and/or radicals (shown by •) involved, as well as arrows showing the movement of electrons. Two types of arrows are used:

 an arrow with a full head (a 'curly arrow') represents a pair of electrons moving

curly arrow

 an arrow with a half arrow-head (a 'fish-hook') represents the movement of a single electron

fish-hook arrow

There are two questions in this exercise. The questions each consist of a central diagram showing the initial stage in a reaction mechanism, surrounded by a selection of suggestions for the result of that step. Your task in each case is to identify which of the diagrams gives the correct outcome of that reaction step. Draw a large arrow showing which diagram is correct, as in the example below.

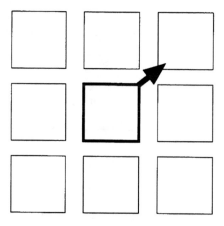

Try and explain your reason(s) for selecting the diagram you chose.

Reaction mechanism 1

I selected this diagram because:

_____

_____

_____

_____

_____

_____

_____

_____

Reaction mechanism 2

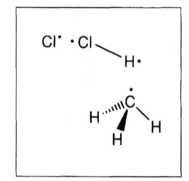

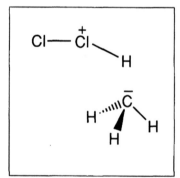

I selected this diagram because:

_____

_____

_____

_____

_____

_____

_____

_____

Reaction mechanisms revealed

Reaction mechanism 1

The diagrams below show and explain the correct answer to the question about the ionic reaction mechanism.

Reaction mechanism 2

The diagrams below show and explain the correct answer to the question about the free radical mechanism.

RS•C

This page has been intentionally left blank.

Scaffolding explanations

Target level

These materials are designed for students taking post-16 chemistry courses.

Topics

The topics covered in these materials include charge: size ratio (charge density); atomic structure and core charge; electronegativity; bond polarity; lattice energy; hydration of ions; hydrogen bonding; melting temperature; boiling temperature; atomic size and base strength.

Rationale

These materials are designed to support students in developing confidence in using explanations in chemistry. In particular the materials aim to demonstrate how a range of phenomena that chemists study are explained in terms of a limited set of basic chemical concepts. The materials are designed to support students by providing frameworks within which they can successfully complete explanations, as a step towards developing explanations on their own. The materials also require students to translate information between a schematic form and standard prose. These notions of 'scaffolding' students' learning, and providing active learning tasks, are discussed in Chapters 3 and 5 of the Teachers' notes.

The materials comprise of two sets of questions (of a type commonly used in public examinations) requiring explanations - which may be used as a pre-test and post-test - a worksheet explaining the key concepts related to these questions, and a worksheet providing structured support in working through the first set of questions.

During piloting it was found some students found the materials helpful, and it was suggested that similar exercises in a wider range of topics would be useful. Some students found the questions difficult, but others found the materials repetitive. This latter response could be seen as a positive outcome, as clearly the students were realising that a few key ideas could be used in a wide range of situations. However, teachers may also wish to use the pre-test to distinguish those students who are already able to use the key concepts to develop satisfactory explanations from those who would benefit from working through the complete set of materials.

Instructions

Each student requires a copy of the following worksheets

■ **Explaining chemical phenomena (1)** – which may be used as a pre-test

■ **Constructing chemical explanations** – which provides a review of the key ideas

■ **Completing explanations** – which provides support in answering the questions on the pre-test

■ **Explaining chemical phenomena (2)** – which may be used as a post-test

Resources

■ Student worksheets
 – Explaining chemical phenomena (1)
 – Constructing chemical explanations

RS•C

- Completing explanations
- Explaining chemical phenomena (2)
- Examples of chemical explanations (1) and (2) – (answers to Explaining chemical phenomena (1) and (2))

Feedback for students

A teachers' answer sheet for **Constructing chemical explanations** is provided. The completed schematics for the questions in the pre- and post- tests are also provided as **Examples of chemical explanations** (1) and (2).

RS•C

RS•C

Scaffolding explanations – answers

Constructing chemical explanations

The table is reproduced below. The answers have been printed in **bold**.

E	Z	n_p	n_e	e.c.	n_{ve}	n_{ce} ($=n_e-n_{ve}$)	n_p-n_{ce} =	c.c.
H	1	1	1	1	1	0	1–0 =	+1
He	2	2	2	2	2	0	2–0 =	+2
Li	3	3	3	2.1	1	2	3–2 =	+1
Be	4	4	4	2.2	2	2	4–2 =	+2
B	5	5	5	2.3	3	2	5–2 =	+3
C	6	6	6	2.4	**4**	2	**6–2 =**	+4
N	7	7	7	**2.5**	5	2	7–2 =	+5
O	8	8	8	2.6	6	2	8–2 =	**+6**
F	9	9	9	2.7	**7**	**2**	**9–2 =**	+7
Ne	10	10	10	2.8	**8**	**2**	10–2 =	**+8**
Na	11	11	11	2.8.1	1	10	11–10 =	+1
Mg	12	**12**	**12**	2.8.2	2	10	12–10 =	+2
Al	13	13	13	2.8.3	3	10	**13–10 =**	+3
Si	14	**14**	**14**	2.8.4	4	**10**	**14–10 =**	+4
P	**15**	**15**	**15**	2.8.5	5	10	**15–10 =**	+5
S	**16**	**16**	**16**	**2.8.6**	6	10	**16–10 =**	+6
Cl	**17**	**17**	**17**	2.8.7	7	**10**	**17–10 =**	+7
Ar	18	18	18	2.8.8	8	10	18–10 =	+8
K	19	**19**	**19**	2.8.8.1	1	18	19–18 =	**+1**
Ca	20	20	20	**2.8.8.2**	2	**18**	20–18 =	+2

Completing the core charge diagram

B: +3 O: +6
P: +5 S: +6
K: +1

(the core charge matches the element's position across the period)

Other answers are provided as photocopiable masters.

Examples of chemical explanations (1)

The following scheme may be used as the basis of explaining the eight questions included in the worksheets **Explaining chemical phenomena** (1).

1. Lithium has a higher melting temperature (454 K) than sodium (371 K).

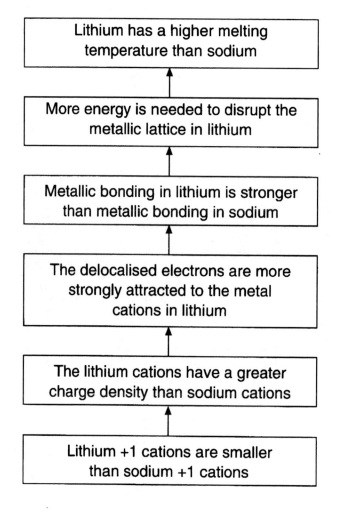

2. There is stronger bonding, called hydrogen bonding, between molecules of water (H_2O) that between molecules of hydrogen sulfide (H_2S).

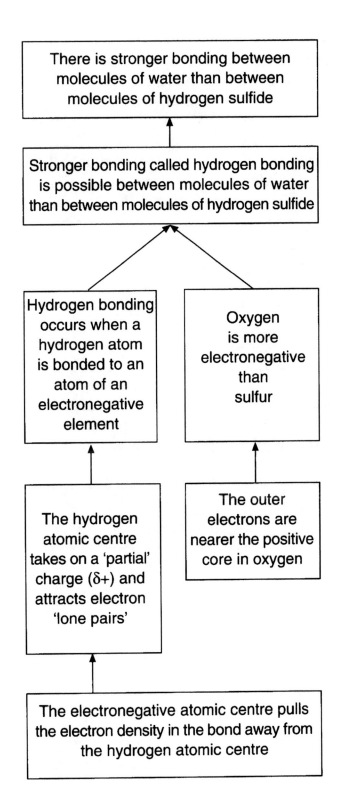

3. The nitrogen atom is smaller than the carbon atom (*ie* it has a smaller covalent radius - 0.074 nm compared to 0.077 nm).

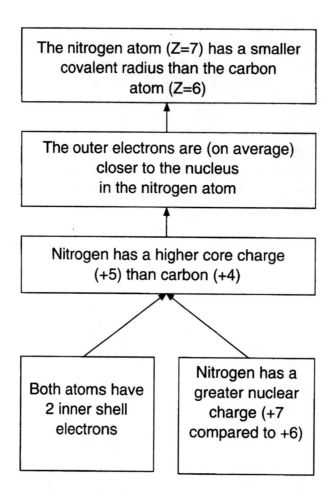

RS•C

4. *Chlorine is more electronegative than bromine.*

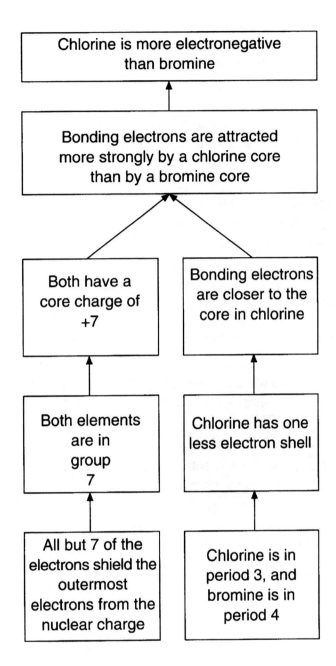

5. Ammonia (NH_3) is a stronger base than phosphine (PH_3).

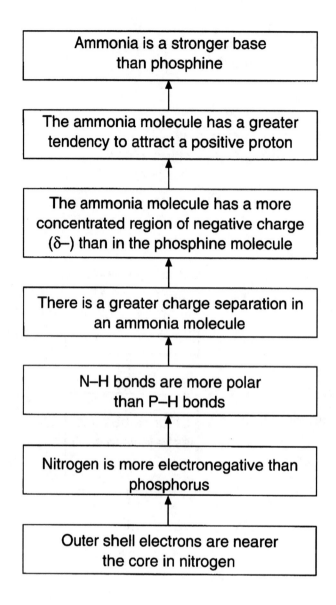

6. Magnesium chloride has a larger lattice energy (2489 kJ mol^{-1}) than calcium chloride (2197 kJmol^{-1}).

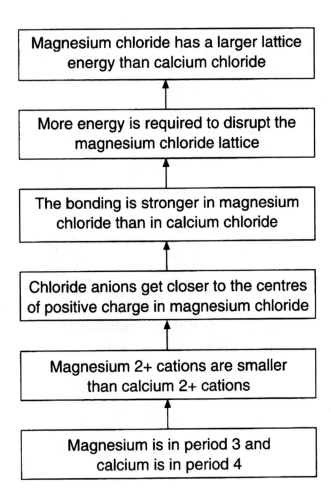

Magnesium chloride has a larger lattice
energy than calcium chloride

↑

More energy is required to disrupt the
magnesium chloride lattice

↑

The bonding is stronger in magnesium
chloride than in calcium chloride

↑

Chloride anions get closer to the centres
of positive charge in magnesium chloride

↑

Magnesium 2+ cations are smaller
than calcium 2+ cations

↑

Magnesium is in period 3 and
calcium is in period 4

7. More energy is released when sodium ions are hydrated (390 kJmol^{-1}) than when potassium ions are hydrated (305 kJmol^{-1}).

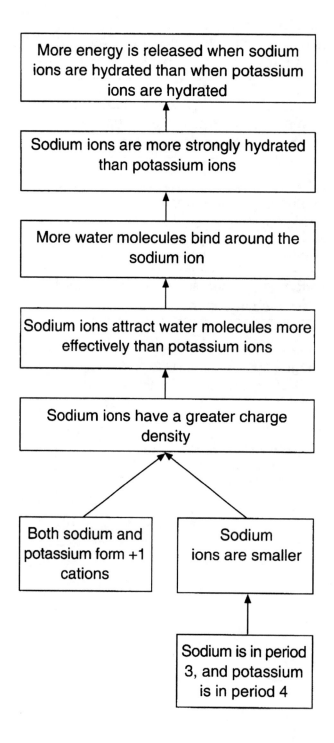

8. Nitrogen is less electronegative than oxygen.

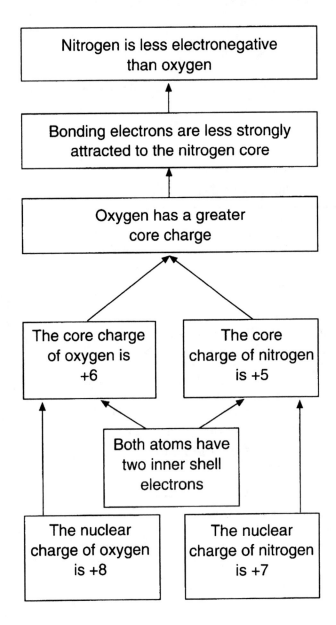

Examples of chemical explanations (2)

The following schemes may be useful as the basis for explaining the eight included in the worksheet **Explaining chemical phenomena (2)**

1. Potassium has a lower melting temperature (336 K) than sodium (371 K).

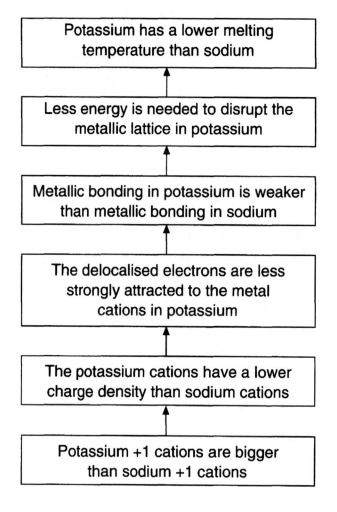

2. There is weaker bonding between molecules of hydrogen chloride (HCl) than between molecules of hydrogen fluoride (HF).

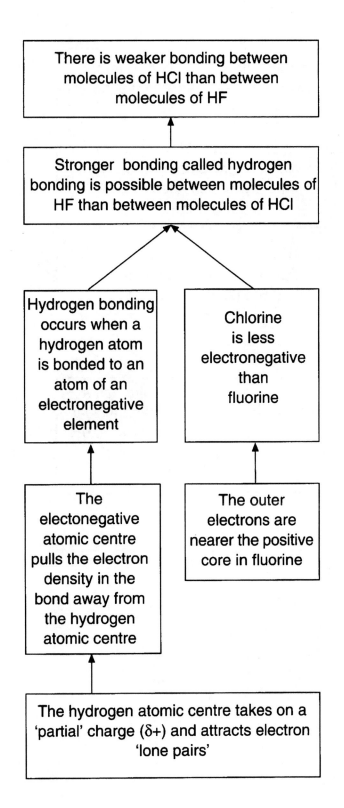

3. The sulfur atom is larger than the chlorine atom (*ie* it has a greater covalent radius – 0.104 nm compared to 0.099 nm).

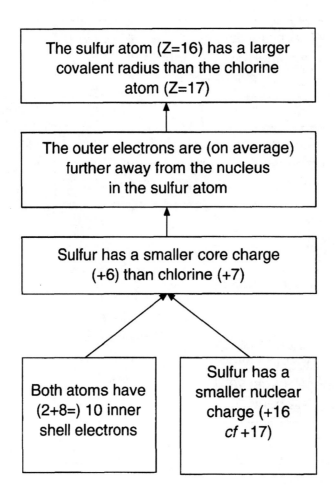

4. Chlorine is less electronegative than fluorine.

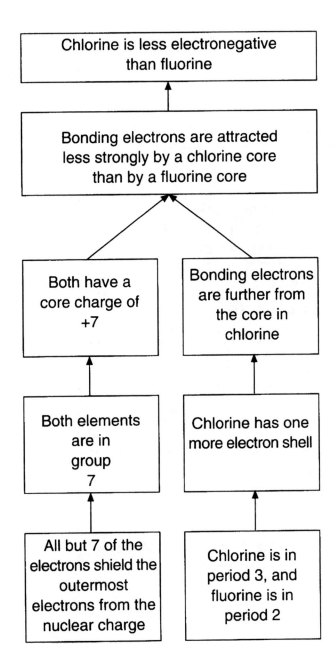

5. Arsine (AsH$_3$) is a weaker base than ammonia (NH$_3$).

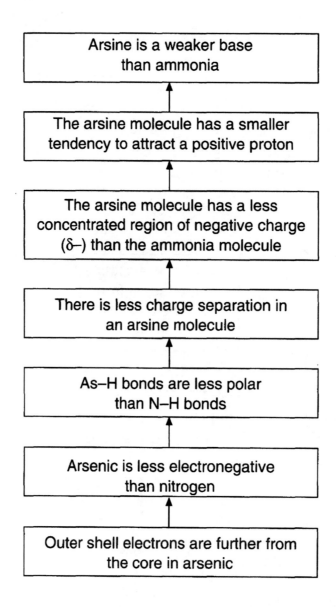

6. Potassium fluoride has a smaller lattice energy (813 kJmol^{-1}) than sodium fluoride (915 kJmol^{-1}).

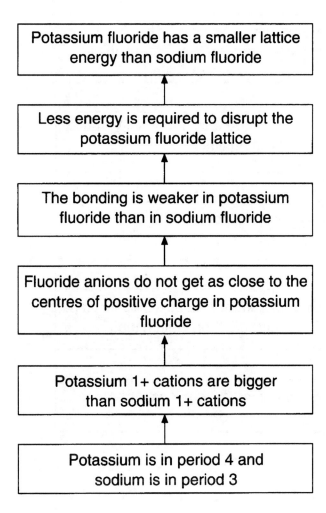

Potassium fluoride has a smaller lattice energy than sodium fluoride

↑

Less energy is required to disrupt the potassium fluoride lattice

↑

The bonding is weaker in potassium fluoride than in sodium fluoride

↑

Fluoride anions do not get as close to the centres of positive charge in potassium fluoride

↑

Potassium 1+ cations are bigger than sodium 1+ cations

↑

Potassium is in period 4 and sodium is in period 3

7. Less energy is released when sodium ions are hydrated (390 kJmol^{-1}) than when lithium ions are hydrated (499 kJmol^{-1}).

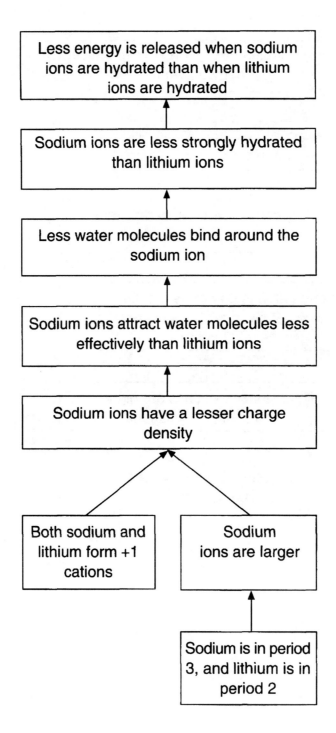

8. Chlorine is more electronegative than sulfur.

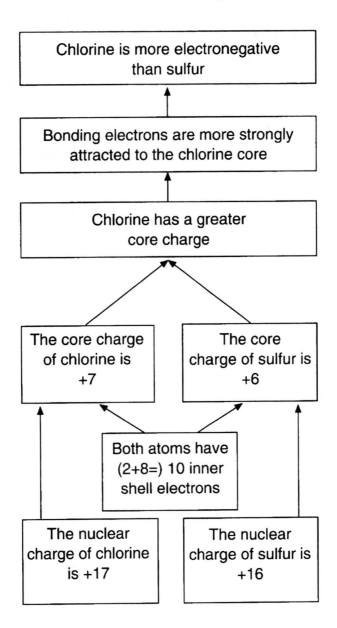

Explaining chemical phenomena (1)

Chemists use their models and theories to try and explain phenomena about chemical systems. Suggest an explanation for each of the following (you may find it useful to refer to a periodic table):

1. Lithium has a higher melting temperature (454 K) than sodium (371 K).

2. There is stronger bonding, called hydrogen bonding, between molecules of water (H_2O) than between molecules of hydrogen sulfide (H_2S).

3. The nitrogen atom is smaller than the carbon atom (*ie* it has a smaller covalent radius – 0.074 nm compared to 0.077 nm).

4. Chlorine is more electronegative than bromine.

5. Ammonia (NH_3) is a stronger base than phosphine (PH_3).

_____

_____

_____

_____

_____

6. Magnesium chloride has a larger lattice energy (2489 $kJmol^{-1}$) than calcium chloride (2197 $kJmol^{-1}$).

_____

_____

_____

_____

_____

7. More energy is released when sodium ions are hydrated (390 $kJmol^{-1}$) than when potassium ions are hydrated (305 $kJmol^{-1}$).

_____

_____

_____

_____

_____

8. Nitrogen is less electronegative than oxygen.

_____

_____

_____

_____

_____

Constructing chemical explanations

Chemists use models and theories to try and explain phenomena about chemical systems. Although chemists seems to use a wide range of different models and theories, many of them are based on the same few basic principles.

If you can learn about these basic ideas you can use them as 'tools' to build up chemical explanations.

Some basic ideas used in explaining chemistry

The importance of size and charge
A large number of chemical phenomena can be explained, at least partly, in terms of simple ideas like the size and charge on ions or other particles.

Charge density
If two ions have the same charge, but are different in size, then the smaller one is said to have a greater density of charge. The ion with the greater density of charge can often have a greater effect – if it can get up close to other ions or molecules.

An ion with greater charge density can form a stronger bond with an oppositely charged ion. This can lead to a more stable ionic lattice, which therefore requires more energy to disrupt. (So the lattice energy of magnesium chloride is greater than the lattice energy of calcium chloride - as the Mg^{2+} ion, which is smaller, has a greater charge density than the Ca^{2+} ion, even though they have the same charge.)

If the ions are of similar size, but have a different charge the one with the greater charge will have the larger charge density, and may be able to form stronger bonds.

When ionic materials dissolve in water the ions are hydrated (surrounded by water molecules which bond to them). The greater the charge density of an ion the more water molecules will bond to it, and the more energy will be released when a material with that type of ion dissolves.

These ideas do not always help us predict what will happen in experiments, as sometimes there are several effects operating at once. For example, an ion with a greater charge density can be hydrated more (which would make the material more soluble) but will usually bond more strongly to oppositely charged ions (which would make the material harder to dissolve!)

In some books the terms charge:size (charge to size ratio) or charge:volume (charge to volume ratio) may be used instead of charge density.

Core charge
Many chemical processes can be – at least partly – explained in terms of the charges in the ions or molecules involved. Atoms are neutral, but separate atoms are seldom involved in chemical processes. Usually we are concerned with ions or molecules.

In what ways are ions and molecules like atoms, and how are they different?

Atoms may be thought of as a positive nucleus surrounded by several shells of electrons. (Of course, the electronic structure is more complicated than that, with different types of orbitals. However, it is often useful to think in terms of shells.) Most of the time the nuclei of the atoms do not change (and when they do this is studied by physicists). Usually only the outermost shell of electrons, the valence shell, is changed in chemical processes. The nucleus and all the inner shells are usually not significantly changed.

The term core is used to describe the nucleus of an atom, plus all the electrons that are not in the outer (valence) shell.

The charge on an atomic core is called the core charge.

It is often useful to know what the core charge is. The core charge will equal the positive charge on the nucleus plus the negative charge of all the inner-shell electrons.

The table shows how to calculate core charge.

Complete the table.

Element E	Atomic Number Z	Number of protons n_p	Number of electrons n_e	Electronic configuration e.c.	Number of outershell electrons n_{ve}	Number of core electrons n_{ce} (=$n_e - n_{ve}$)	To calculate core charge $n_p - n_{ce}$ =	Core charge c.c.
H	1	1	1	1	1	0	1-0 =	+1
He	2	2	2	2	2	0	2-0 =	+2
Li	3	3	3	2.1	1	2	3-2 =	+1
Be	4	4	4	2.2	2	2	4-2 =	+2
B	5	5	5	2.3	3	2	5-2 =	+3
C	6	6	6	2.4		2		+4
N	7	7	7		5	2	7-2 =	+5
O	8	8	8	2.6	6	2	8-2 =	
F	9	9	9	2.7				+7
Ne	10	10	10	2.8			10-2 =	
Na	11	11	11	2.8.1	1	10	11-10 =	+1
Mg				2.8.2	2	10	12-10 =	+2
Al	13	13	13	2.8.3		10		+3
Si	14			2.8.4	4			+4
P				2.8.5		10		
S					6	10		
Cl				2.8.7				+7
Ar	18	18	18	2.8.8	8	10	18-10 =	+8
K	19			2.8.8.1	1	18	19-18 =	
Ca	20	20	20		2			+2

The diagram below shows the symbols for the first 20 elements. Using the table on the previous page, complete the diagram by adding the core charges that are not shown.

Can you spot any pattern?

H +1	Core charge for the first 20 elements						He +2
Li +1	Be +2	B	C +4	N +5	O	F +7	Ne +8
Na +1	Mg +2	Al +3	Si +4	P	S	Cl +7	Ar +8
K	Ca +2						

Electronegativity

One of the most useful chemical concepts is that of electronegativity.

Electronegativity is the tendency of an element to attract the bonding electrons towards itself in compounds.

The electronegativity of an element depends upon how strongly the outer (valence) shell electrons are attracted to the core. The greater the charge on the core, and the nearer the outer shell of electrons are, the more strongly they are bound to the core.

Electronegativity is therefore greater at the top of a group (where the outer shell of negatively charged electrons is pulled closest to the positively charged nucleus) and to the right of a period (where the core charge is greatest):

H (+1)	Approximate core charge density for some of the elements					
Li (+1)	Be (+2)	B (+3)	C (+4)	N (+5)	O (+6)	F (+7)
Na (+1)	Mg (+2)	Al (+3)	Si (+4)	P (+5)	S (+6)	Cl (+7)
K (+1)	Ca (+2)					

Bond polarity in terms of electronegativity

Where the electrons in a bond are pulled equally by both cores, the bond is non-polar.

For example in F_2 both the atoms of fluorine have cores of the same size and charge.

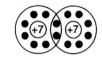

However, in the molecule of CF_4 the fluorine atoms have a more highly charged core , and the bonding electrons are pulled closer to the fluorine cores than carbon core, The bonds are polar. This can be represented as:

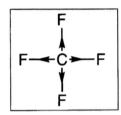

The inter-halogen compound ClF also has a polar bond. Although the cores of both atoms have the same charge, the fluorine core has a higher charge density. This can be represented as:

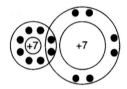

δ+   δ–

Cl —— F

Building up chemistry explanations

Many of the explanations we construct in chemistry use ideas such as charge density, core charge and electronegativity. Many explanations can be built up using just a few basis ideas. Certain key phrases can be used as tools for building up explanations. (This does not mean you can just pick any key phrase: you have to understand the chemistry, and select the right phrases for a particular explanation!)

Consider the question: why do water molecules bond together?

Water molecules bond together because they are polar.

Water molecules are polar because the hydrogen-oxygen bond is polar.

The hydrogen-oxygen bond is polar because oxygen is more electronegative than hydrogen.

Oxygen is more electronegative than hydrogen because oxygen has a larger core charge.

This is not a 'complete' explanation. It would be possible to explain in more detail about the core charge (eg oxygen has a core charge of +6 because it has a nucleus of charge +8 , partly shielded by 2 electrons in the core). It is also important to know that a molecule with polar bonds is not always a polar molecule! (So the molecule of CF_4 discussed above has four polar bonds, but overall the molecule is non-polar.) The explanation above could be improved by adding:

... Water molecules are polar because the hydrogen-oxygen bond is polar, and the water molecule has two polar bonds which do not cancel out ...

This theme could also be continued...

... The two polar bonds do not cancel because the water molecule is angular.

The water molecule is angular because there are four electron pairs in the oxygen outer shell.

There are two non-bonding (lone) pairs as well as the two pairs of bonding electrons ...

So even a simple question could have a very complicated answer if we want to give a full and detailed explanation. You do not normally need to do this, although it can be good practice in testing how well you understand your chemistry.

Charting your explanations

If some explanations seem rather complicated, you may find it helps to break the explanation down into steps, which can be put into a flow chart. In the example below, the phenomenon to be explained is written at the top of the box, and each arrow may be read as a 'because'.

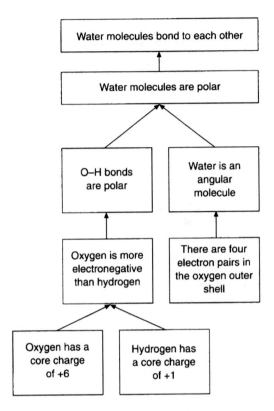

Although such a scheme may seem complicated, it helps to learn to think about the individual steps in the explanations. If you can understand the steps you can learn to put together such schemes.

For example consider the question: why does water have a higher boiling temperature than would be predicted from its molecular mass?

You may know that it is because the hydrogen bonding in water holds the molecules together, so that more energy is needed to separate the molecules. In other words, we can amend the scheme above.

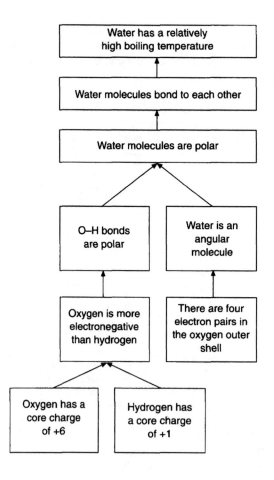

Some key phrases

The following phrases are examples of some of those that may be useful in constructing explanations (you may want to add to this list):

- has strong(er)/weak(er) bonding
- is more/less electronegative
- has a higher/lower charge density
- has a larger/smaller core charge
- electrons are nearer/further from the nucleus

RS•C

Completing explanations

1. Explaining that lithium has a higher melting temperature than sodium

Complete the missing words in the scheme (left).

In your own words try and explain why lithium has a higher melting temperature (454 K) than sodium (371 K).

_____

_____

_____

_____

_____

_____

_____

_____

_____

_____

_____

_____

_____

_____

_____

Can you redraw the scheme to include extra detail, and so make it more informative?

Lithium has a higher melting temperature than sodium

↑

More energy is needed to disrupt the metallic lattice in _____

↑

Metallic bonding in lithium is stronger than metallic bonding in sodium

↑

The delocalised electrons are more strongly attracted to the metal cations in lithium

↑

The lithium cations have a _____ charge density than sodium cations

↑

Lithium +1 cations are smaller than sodium +1 cations

Completing explanations

2. Explaining that there is stronger bonding, called hydrogen bonding, between molecules of water (H_2O) than between molecules of hydrogen sulfide (H_2S)

Complete the missing words in the scheme (left).

In your own words try and explain why there is stronger bonding called hydrogen bonding between molecules of water (H_2O) than between molecules of hydrogen sulfide (H_2S).

_____

_____

_____

_____

_____

_____

_____

_____

_____

_____

_____

_____

_____

Can you redraw the scheme to include extra detail, and so make it more informative?

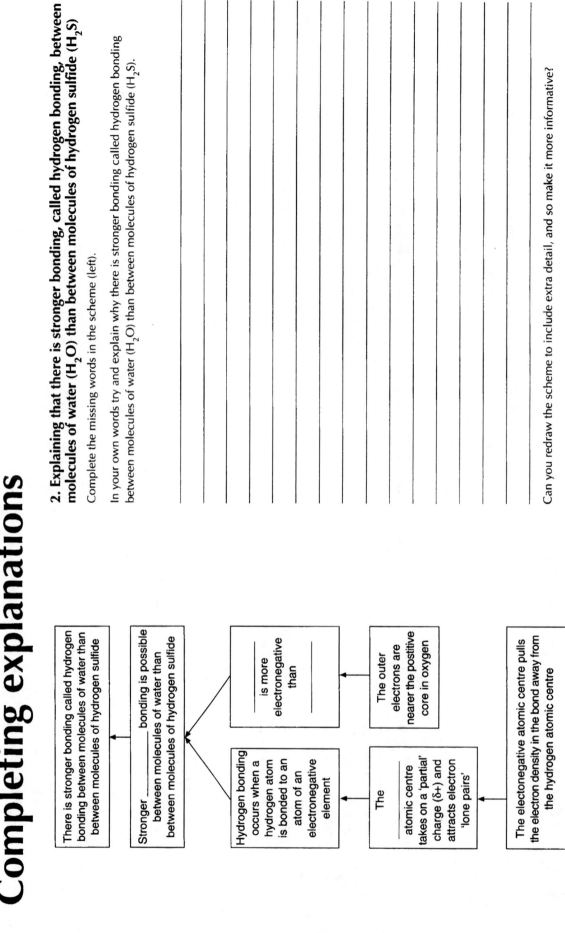

There is stronger bonding called hydrogen bonding between molecules of water than between molecules of hydrogen sulfide

Stronger _____ bonding is possible between molecules of water than between molecules of hydrogen sulfide

_____ is more electronegative than _____

Hydrogen bonding occurs when a hydrogen atom is bonded to an atom of an electronegative element

The outer electrons are nearer the positive core in oxygen

The _____ atomic centre takes on a 'partial' charge (δ+) and attracts electron 'lone pairs'

The electonegative atomic centre pulls the electron density in the bond away from the hydrogen atomic centre

Completing explanations

3. Explaining that the nitrogen atom is smaller than the carbon atom (*ie* it has a smaller covalent radius – 0.074 nm compared to 0.077 nm)

Complete the missing words in the scheme (left).

In your own words try and explain why the nitrogen atom is smaller than the carbon atom.

_____

_____

_____

_____

_____

_____

_____

_____

_____

_____

_____

_____

_____

_____

_____

Can you redraw the scheme to include extra detail, and so make it more informative?

The nitrogen atom (Z=7) has a smaller covalent radius than the carbon atom (Z=6)

The outer electrons are (on average) _____ the nucleus in the nitrogen atom

Nitrogen has a higher core charge (+5) than carbon (+4)

Nitrogen has a greater _____ charge (+7 compared to +6)

Both atoms have _____ inner shell electrons

Completing explanations

4. Explaining that chlorine is more electronegative than bromine

Complete the missing words in the scheme (left).

In your own words try and explain why chlorine is more electronegative than bromine.

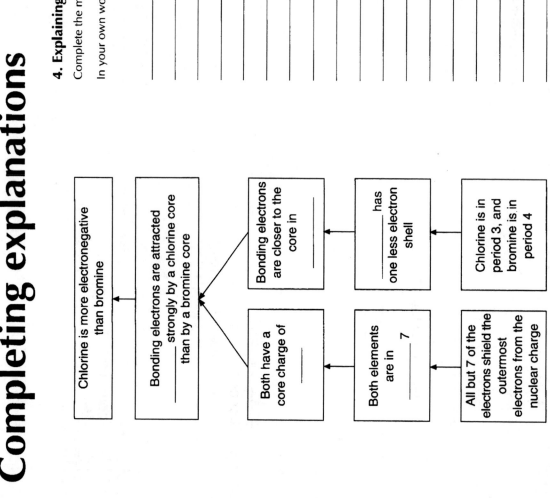

Chlorine is more electronegative than bromine

Bonding electrons are attracted _____ strongly by a chlorine core than by a bromine core

Bonding electrons are closer to the core in _____

_____ has one less electron shell

Chlorine is in period 3, and bromine is in period 4

Both have a core charge of _____

Both elements are in _____ 7

All but 7 of the electrons shield the outermost electrons from the nuclear charge

Can you redraw the scheme to include extra detail, and so make it more informative?

Completing explanations

5. Explaining that ammonia (NH_3) is a stronger base than phosphine (PH_3)

Complete the missing words in the scheme (left).

In your own words try and explain why ammonia (NH_3) is a stronger base than phosphine (PH_3).

_____
_____
_____
_____
_____
_____
_____
_____
_____
_____
_____
_____
_____
_____
_____

Can you redraw the scheme to include extra detail, and so make it more informative?

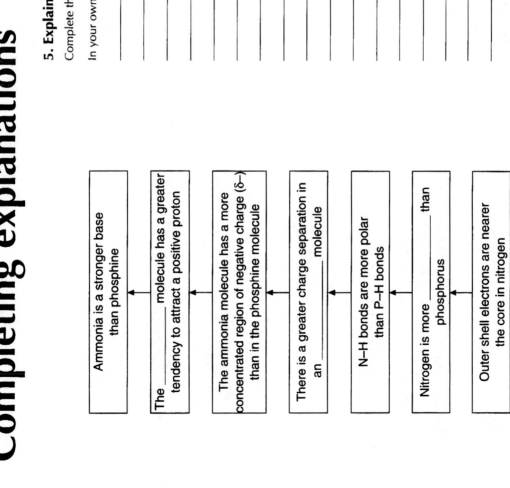

Ammonia is a stronger base than phosphine

↑

The _____ molecule has a greater tendency to attract a positive proton

↑

The ammonia molecule has a more concentrated region of negative charge (δ−) than in the phosphine molecule

↑

There is a greater charge separation in an _____ molecule

↑

N–H bonds are more polar than P–H bonds

↑

Nitrogen is more _____ than phosphorus

↑

Outer shell electrons are nearer the core in nitrogen

Completing explanations

6. Explaining that magnesium chloride has a larger lattice energy (2489 kJmol^{-1}) than calcium chloride (2197 kJmol^{-1})

Complete the missing words in the scheme (left).

In your own words try and explain why magnesium chloride has a larger lattice energy than calcium chloride.

_____

_____

_____

_____

_____

_____

_____

_____

_____

_____

_____

_____

Can you redraw the scheme to include extra detail, and so make it more informative?

| Magnesium chloride has a larger lattice energy than calcium chloride |

↑

| More energy is required to disrupt the _____ chloride lattice |

↑

| The bonding is stronger in magnesium chloride than in calcium chloride |

↑

| Chloride anions get closer to the centres of positive charge in _____ chloride |

↑

| Magnesium 2+ cations are _____ than calcium 2+ cations |

↑

| Magnesium is in period 3 and calcium is in period 4 |

RS•C

Completing explanations

7. Explaining that more energy is released when sodium ions are hydrated (390 kJmol^{-1}) than when potassium ions are hydrated (305 kJmol^{-1})

Complete the missing words in the scheme (left).

In your own words try and explain why more energy is released when sodium ions are hydrated than when potassium ions are hydrated.

_____

_____

_____

_____

_____

_____

_____

_____

_____

_____

_____

Can you redraw the scheme to include extra detail, and so make it more informative?

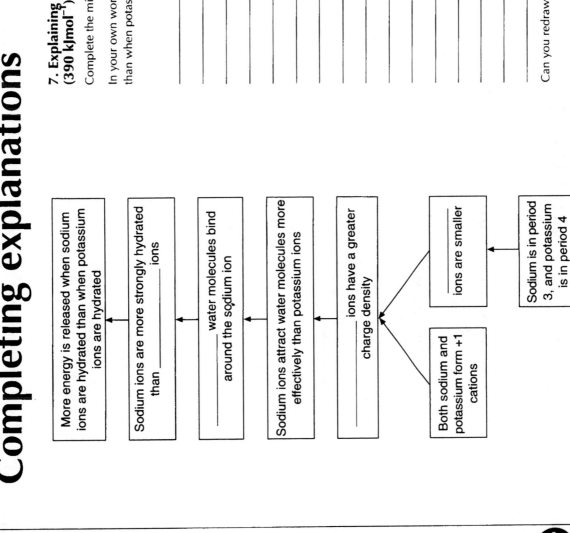

More energy is released when sodium ions are hydrated than when potassium ions are hydrated

↑

Sodium ions are more strongly hydrated than _____ ions

↑

_____ water molecules bind around the sodium ion

↑

Sodium ions attract water molecules more effectively than potassium ions

↑

_____ ions have a greater charge density

↑

_____ ions are smaller

↑

Sodium is in period 3, and potassium is in period 4

Both sodium and potassium form +1 cations

RS•C

Completing explanations

8. Explaining that nitrogen is less electronegative than oxygen.

Complete the missing words in the scheme (left).

In your own words try and explain why nitrogen is less electronegative than oxygen.

_____

_____

_____

_____

_____

_____

_____

_____

_____

_____

_____

_____

_____

_____

_____

Can you redraw the scheme to include extra detail, and so make it more informative?

Nitrogen is less electronegative than oxygen

Bonding electrons are less strongly attracted to the _____ core

_____ has a greater _____ core charge

The _____ charge of nitrogen is +5

The core charge of oxygen is _____

Both atoms have two inner shell electrons

The nuclear charge of nitrogen is _____

The _____ charge of oxygen is +8

RS•C

Explaining chemical phenomena (2)

Chemists use their models and theories to try and explain phenomena about chemical systems. Suggest an explanation for each of the following (you may find it useful to refer to a Periodic Table).

1. Potassium has a lower melting temperature (336 K) than sodium (371 K).

_____

_____

_____

_____

_____

2. There is weaker bonding between molecules of hydrogen chloride (HCl) that between molecules of hydrogen fluoride (HF).

_____

_____

_____

_____

_____

3. The sulfur atom is larger than the chlorine atom (_ie_ it has a greater covalent radius – 0.104 pm compared to 0.099 pm).

_____

_____

_____

_____

_____

4. Chlorine is less electronegative than fluorine.

_____

_____

_____

_____

_____

5. Arsine (AsH_3) is a weaker base than ammonia (NH_3).

_____

_____

_____

_____

_____

6. Potassium fluoride has a smaller lattice energy (813 $kJmol^{-1}$) than sodium fluoride (915 $kJmol^{-1}$).

_____

_____

_____

_____

_____

7. Less energy is released when sodium ions are hydrated (390 $kJmol^{-1}$) than when lithium ions are hydrated (499 $kJmol^{-1}$).

_____

_____

_____

_____

_____

8. Chlorine is more electronegative than sulfur.

_____

_____

_____

_____

_____

RS•C

RS•C

Chemical comparisons

Target level

This resource is a set of probes for use across the 11–19 age range.

Topic

These materials are designed to be open-ended. Although individual probes may be selected to match teaching topics, the materials may also be used to as a way of exploring the extent of students' ideas at various stages of a course.

Rationale

The materials ask students to suggest similarities and differences between various pairs of chemical systems presented as simple diagrams. This is one way of exploring a student's repertoire of chemical ideas - and although the focus is on what students can suggest, the exercise can provide evidence of glaring omissions as well as of alternative conceptions. The activity provides an opportunity for students to undertake a task with no single 'right' answer (although teachers will identify certain responses they would expect students to make at different ages), and thus provides scope for students to demonstrate creativity and lateral thinking. Students are also given the opportunity to indicate which responses they feel are important to chemistry to allow them to distinguish significant from trivial comparisons.

These ideas are discussed in Chapter 2 of the Teachers' notes.

Teachers may find their students providing only one or two responses. This is discussed in Chapter 7 of the Teachers' notes.

During piloting it was found that the probes revealed deeply held alternative conceptions, and revealed areas where students were confused. Students tended to find the activity challenging (perhaps indicating that students are more familiar with closed questions in science), but it provided a basis for valuable discussion. The activity was considered useful.

Instructions

The teacher should select one or more probes suitable for the age and knowledge base of the group. The probes may be set as individual work, or as the focus for group discussion. Probes 1–20 may be more suitable for pre-16 work.

Resources

■ Student worksheets
  – Chemical comparisons

The worksheets invite comparison between:

1. Iron and sulfur

2. Water and sugar solution

3. Sea water and mercury

4. Particle models of gas and liquid

5. Particle model of liquid and particle model of solid

6. Salt crystal and sulfur crystal

RS•C

7. Ionic models of silver chloride and sodium chloride

8. Particle models of solid sodium chloride and sodium chloride solution

9. Bonding in carbon (diamond) and sulfur

10. Structural formulae of ethanoic acid and carbon dioxide

11. Formulae of sulfur and copper(II) sulfate

12. Electron shell model of fluorine and formula of hydrogen fluoride

13. Particle models of water and helium

14. Particle models of iron and sodium chloride

15. Models of bonding in carbon (diamond) and silver nitrate

16. Electron shell models of a sodium atom and a fluorine atom

17. Electron shell models of a hydrogen molecule and a hydrogen atom

18. Particle models of solid iodine and nitrogen gas

19. Particle models of iron and copper

20. Particle models of iodine and magnesium oxide

21. Electron cloud and electron shell models

22. Structural formula and electron cloud models

23. Different structural formulae (inorganic)

24. Different structural formulae (organic)

25. Electron cloud and structural formulae

26. Electron pair and free-radical mechanisms

27. Two electron cloud models

28. Reaction profiles for endothermic and exothermic reactions

29. Structural formulae of ketone and alkenal

30. Stereochemistry of PF_5 and $CuCl_4^{2-}$

31. Bonding of CF_4 and sterochemistry of CH_4

32. Structural formulae of halogenoalkanes

33. Electron shell diagrams for CF_4 and F_2

34. Displayed formulae for benzene and cyclohexene

35. Electron cloud diagrams for NH_3 and BCl_3

36. Structural formulae of ethanoic acid and ethanoic acid dimer

Feedback for students

It may be useful to make OHP transparency copies of the probes used as a basis for discussing the comparisons with the class. Discussion should be based around sharing ideas rather than simply judging responses are correct or incorrect. It is suggested that all valid comparisons should be encouraged, but that those relating to key curriculum ideas should be emphasised.

Chemical comparisons

The two diagrams below show things you might study in chemistry. Think about how the things shown in the diagrams are similar and how they are different:

Iron

Sulfur

List the similarities and differences you can think of below.

In which ways are they alike?

_____

_____

_____

_____

_____

_____

_____

_____

In which ways are they different?

_____

_____

_____

_____

_____

_____

_____

_____

Which of these similarities and differences do you think are important to chemists? Put a star symbol (*) in front of the important similarities, and the important differences.

Chemical comparisons

The two diagrams below show things you might study in chemistry. Think about how the things shown in the diagrams are similar and how they are different:

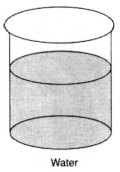

Water

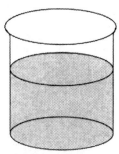

Sugar solution

List the similarities and differences you can think of below.

In which ways are they alike?

_____

_____

_____

_____

_____

_____

_____

_____

In which ways are they different?

_____

_____

_____

_____

_____

_____

_____

Which of these similarities and differences do you think are important to chemists? Put a star symbol (*) in front of the important similarities, and the important differences.

Chemical comparisons

The two diagrams below show things you might study in chemistry. Think about how the things shown in the diagrams are similar and how they are different:

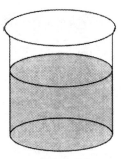

Sea water

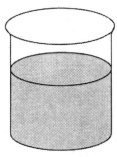

Mercury

List the similarities and differences you can think of below.

In which ways are they alike?

_____

_____

_____

_____

_____

_____

_____

_____

In which ways are they different?

_____

_____

_____

_____

_____

_____

_____

Which of these similarities and differences do you think are important to chemists? Put a star symbol (*) in front of the important similarities, and the important differences.

Chemical comparisons

The two diagrams below show things you might study in chemistry. Think about how the things shown in the diagrams are similar and how they are different:

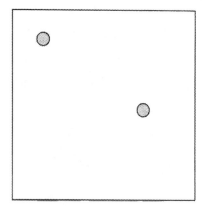
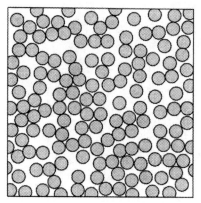

List the similarities and differences you can think of below.

In which ways are they alike?

_____

_____

_____

_____

_____

_____

_____

_____

In which ways are they different?

_____

_____

_____

_____

_____

_____

_____

Which of these similarities and differences do you think are important to chemists? Put a star symbol (*) in front of the important similarities, and the important differences.

RS•C

Chemical comparisons

The two diagrams below show things you might study in chemistry. Think about how the things shown in the diagrams are similar and how they are different:

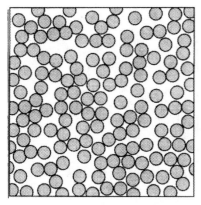 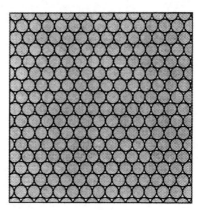

List the similarities and differences you can think of below.

In which ways are they alike?

_____

_____

_____

_____

_____

_____

_____

_____

In which ways are they different?

_____

_____

_____

_____

_____

_____

_____

_____

Which of these similarities and differences do you think are important to chemists? Put a star symbol (*) in front of the important similarities, and the important differences.

Chemical comparisons

The two diagrams below show things you might study in chemistry. Think about how the things shown in the diagrams are similar and how they are different:

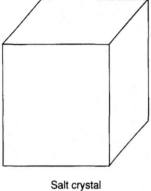

Salt crystal

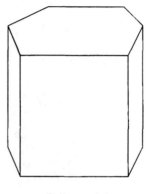

Sulfur crystal

List the similarities and differences you can think of below.

In which ways are they alike?

_____

_____

_____

_____

_____

_____

_____

_____

In which ways are they different?

_____

_____

_____

_____

_____

_____

_____

Which of these similarities and differences do you think are important to chemists? Put a star symbol (*) in front of the important similarities, and the important differences.

RS•C

Chemical comparisons

The two diagrams below show things you might study in chemistry. Think about how the things shown in the diagrams are similar and how they are different:

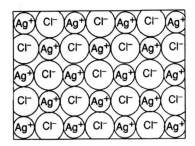   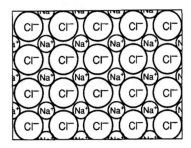

List the similarities and differences you can think of below.

In which ways are they alike?

_____

_____

_____

_____

_____

_____

_____

In which ways are they different?

_____

_____

_____

_____

_____

_____

_____

_____

Which of these similarities and differences do you think are important to chemists? Put a star symbol (*) in front of the important similarities, and the important differences.

Chemical comparisons

The two diagrams below show things you might study in chemistry. Think about how the things shown in the diagrams are similar and how they are different:

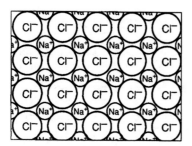
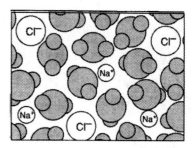

List the similarities and differences you can think of below.

In which ways are they alike?

_____

_____

_____

_____

_____

_____

_____

_____

In which ways are they different?

_____

_____

_____

_____

_____

_____

_____

_____

Which of these similarities and differences do you think are important to chemists? Put a star symbol (*) in front of the important similarities, and the important differences.

Chemical comparisons

The two diagrams below show things you might study in chemistry. Think about how the things shown in the diagrams are similar and how they are different:

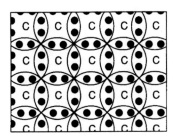

List the similarities and differences you can think of below.

In which ways are they alike?

_____

_____

_____

_____

_____

_____

_____

_____

In which ways are they different?

_____

_____

_____

_____

_____

_____

_____

_____

Which of these similarities and differences do you think are important to chemists? Put a star symbol (*) in front of the important similarities, and the important differences.

Chemical comparisons

The two diagrams below show things you might study in chemistry. Think about how the things shown in the diagrams are similar and how they are different:

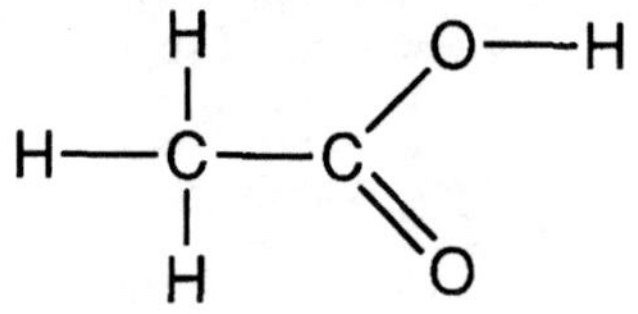

$$H-\overset{\overset{\displaystyle H}{|}}{\underset{\underset{\displaystyle H}{|}}{C}}-\overset{\overset{\displaystyle O-H}{}}{\underset{\underset{\displaystyle O}{\|}}{C}}$$

$$O = C = O$$

List the similarities and differences you can think of below.

In which ways are they alike?

_____

_____

_____

_____

_____

_____

_____

_____

In which ways are they different?

_____

_____

_____

_____

_____

_____

_____

Which of these similarities and differences do you think are important to chemists? Put a star symbol (*) in front of the important similarities, and the important differences.

RS•C

Chemical comparisons

The two diagrams below show things you might study in chemistry. Think about how the things shown in the diagrams are similar and how they are different:

$$S_8 \qquad\qquad CuSO_4$$

List the similarities and differences you can think of below.

In which ways are they alike?

_____

_____

_____

_____

_____

_____

_____

In which ways are they different?

_____

_____

_____

_____

_____

_____

Which of these similarities and differences do you think are important to chemists? Put a star symbol (*) in front of the important similarities, and the important differences.

Chemical comparisons

The two diagrams below show things you might study in chemistry. Think about how the things shown in the diagrams are similar and how they are different:

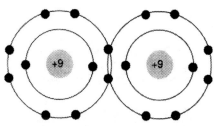

Fluorine molecule

List the similarities and differences you can think of below.

In which ways are they alike?

_____
_____
_____
_____
_____
_____
_____
_____

In which ways are they different?

_____
_____
_____
_____
_____
_____
_____
_____

Which of these similarities and differences do you think are important to chemists? Put a star symbol (*) in front of the important similarities, and the important differences.

RS•C

Chemical comparisons

The two diagrams below show things you might study in chemistry. Think about how the things shown in the diagrams are similar and how they are different:

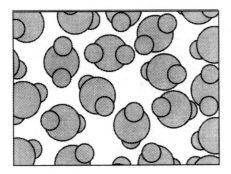
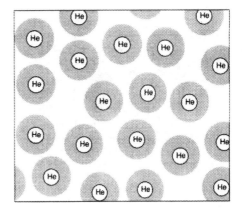

List the similarities and differences you can think of in the tables below.

In which ways are they alike?

_____

_____

_____

_____

_____

_____

_____

_____

In which ways are they different?

_____

_____

_____

_____

_____

_____

_____

Which of these similarities and differences do you think are important to chemists? Put a star symbol (*) in front of the important similarities, and the important differences.

Chemical comparisons

The two diagrams below show things you might study in chemistry. Think about how the things shown in the diagrams are similar and how they are different:

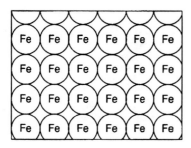

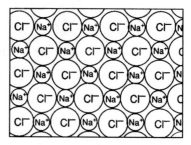

List the similarities and differences you can think of below.

In which ways are they alike?

_____

_____

_____

_____

_____

_____

_____

_____

In which ways are they different?

_____

_____

_____

_____

_____

_____

_____

Which of these similarities and differences do you think are important to chemists? Put a star symbol (*) in front of the important similarities, and the important differences.

RS•C

Chemical comparisons

The two diagrams below show things you might study in chemistry. Think about how the things shown in the diagrams are similar and how they are different:

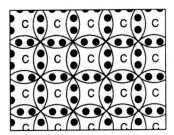   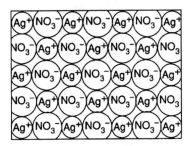

List the similarities and differences you can think of below.

In which ways are they alike?

_____

_____

_____

_____

_____

_____

_____

_____

In which ways are they different?

_____

_____

_____

_____

_____

_____

_____

_____

Which of these similarities and differences do you think are important to chemists? Put a star symbol (*) in front of the important similarities, and the important differences.

Chemical comparisons

The two diagrams below show things you might study in chemistry. Think about how the things shown in the diagrams are similar and how they are different:

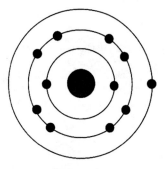

Sodium atom

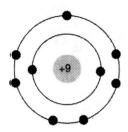

Fluorine atom

List the similarities and differences you can think of below.

In which ways are they alike?

_____

_____

_____

_____

_____

_____

_____

_____

In which ways are they different?

_____

_____

_____

_____

_____

_____

_____

_____

Which of these similarities and differences do you think are important to chemists? Put a star symbol (*) in front of the important similarities, and the important differences.

RS•C

Chemical comparisons

The two diagrams below show things you might study in chemistry. Think about how the things shown in the diagrams are similar and how they are different:

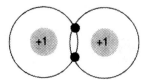

Hydrogen molecule

Hydrogen atom

List the similarities and differences you can think of below.

In which ways are they alike?

_____

_____

_____

_____

_____

_____

_____

In which ways are they different?

_____

_____

_____

_____

_____

_____

_____

Which of these similarities and differences do you think are important to chemists? Put a star symbol (*) in front of the important similarities, and the important differences.

Chemical comparisons

The two diagrams below show things you might study in chemistry. Think about how the things shown in the diagrams are similar and how they are different:

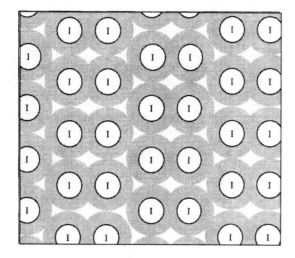
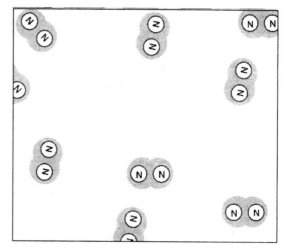

List the similarities and differences you can think of below.

In which ways are they alike?

_____

_____

_____

_____

_____

_____

_____

_____

In which ways are they different?

_____

_____

_____

_____

_____

_____

_____

_____

Which of these similarities and differences do you think are important to chemists? Put a star symbol (*) in front of the important similarities, and the important differences.

RS•C

Chemical comparisons

The two diagrams below show things you might study in chemistry. Think about how the things shown in the diagrams are similar and how they are different:

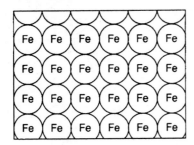
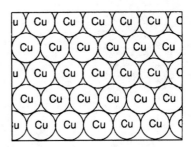

List the similarities and differences you can think of below.

In which ways are they alike?

_____

_____

_____

_____

_____

_____

_____

In which ways are they different?

_____

_____

_____

_____

_____

_____

_____

Which of these similarities and differences do you think are important to chemists? Put a star symbol (*) in front of the important similarities, and the important differences.

Chemical comparisons

The two diagrams below show things you might study in chemistry. Think about how the things shown in the diagrams are similar and how they are different:

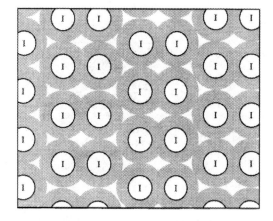
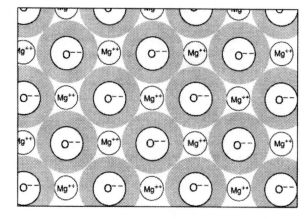

List the similarities and differences you can think of below.

In which ways are they alike?

_____

_____

_____

_____

_____

_____

_____

_____

In which ways are they different?

_____

_____

_____

_____

_____

_____

_____

Which of these similarities and differences do you think are important to chemists? Put a star symbol (*) in front of the important similarities, and the important differences.

Chemical comparisons

The two diagrams below show things you might study in chemistry. Think about how the things shown in the diagrams are similar and how they are different:

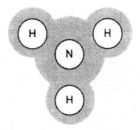

Ammonia molecule

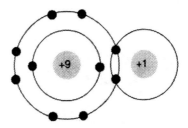

Hydrogen fluoride molecule

List the similarities and differences you can think of below.

In which ways are they alike?

_____

_____

_____

_____

_____

_____

_____

In which ways are they different?

_____

_____

_____

_____

_____

_____

_____

Which of these similarities and differences do you think are important to chemists? Put a star symbol (*) in front of the important similarities, and the important differences.

Chemical comparisons

The two diagrams below show things you might study in chemistry. Think about how the things shown in the diagrams are similar and how they are different:

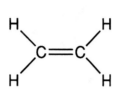

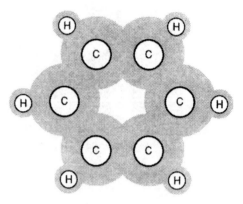

List the similarities and differences you can think of below.

In which ways are they alike?

_____

_____

_____

_____

_____

_____

_____

In which ways are they different?

_____

_____

_____

_____

_____

_____

_____

Which of these similarities and differences do you think are important to chemists? Put a star symbol (*) in front of the important similarities, and the important differences.

Chemical comparisons

The two diagrams below show things you might study in chemistry. Think about how the things shown in the diagrams are similar and how they are different:

H—Cl

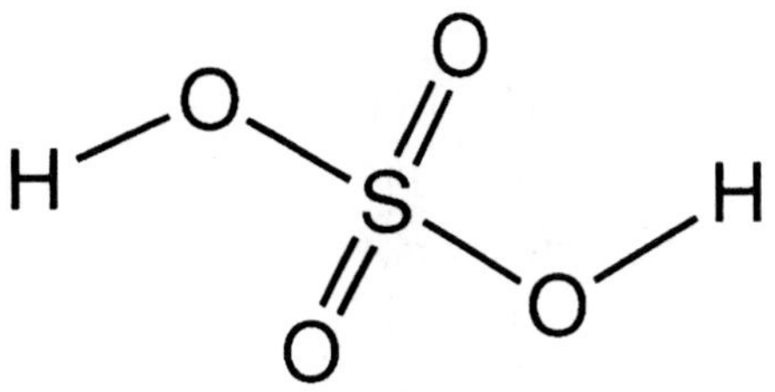

List the similarities and differences you can think of below.

In which ways are they alike?

_____

_____

_____

_____

_____

_____

_____

_____

In which ways are they different?

_____

_____

_____

_____

_____

_____

_____

_____

Which of these similarities and differences do you think are important to chemists? Put a star symbol (*) in front of the important similarities, and the important differences.

Chemical comparisons

The two diagrams below show things you might study in chemistry. Think about how the things shown in the diagrams are similar and how they are different:

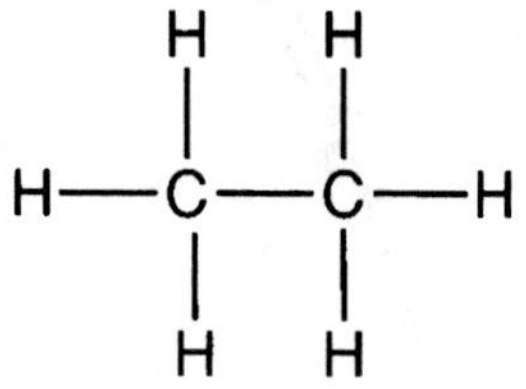

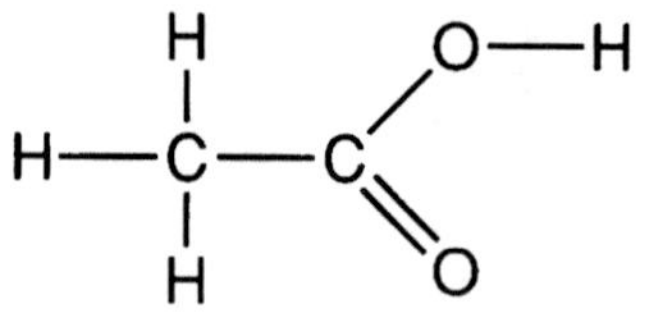

List the similarities and differences you can think of below.

In which ways are they alike?

_____

_____

_____

_____

_____

_____

_____

In which ways are they different?

_____

_____

_____

_____

_____

_____

_____

Which of these similarities and differences do you think are important to chemists? Put a star symbol (*) in front of the important similarities, and the important differences.

RS•C

Chemical comparisons

The two diagrams below show things you might study in chemistry. Think about how the things shown in the diagrams are similar and how they are different:

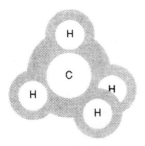

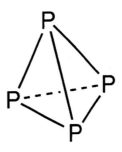

List the similarities and differences you can think of below.

In which ways are they alike?

_____

_____

_____

_____

_____

_____

_____

_____

In which ways are they different?

_____

_____

_____

_____

_____

_____

_____

Which of these similarities and differences do you think are important to chemists? Put a star symbol (*) in front of the important similarities, and the important differences.

Chemical comparisons

The two diagrams below show things you might study in chemistry. Think about how the things shown in the diagrams are similar and how they are different:

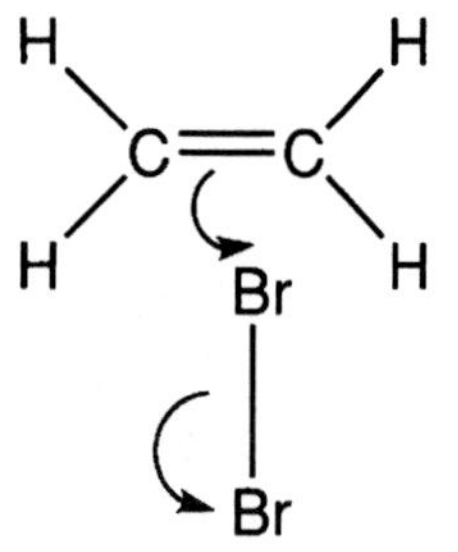

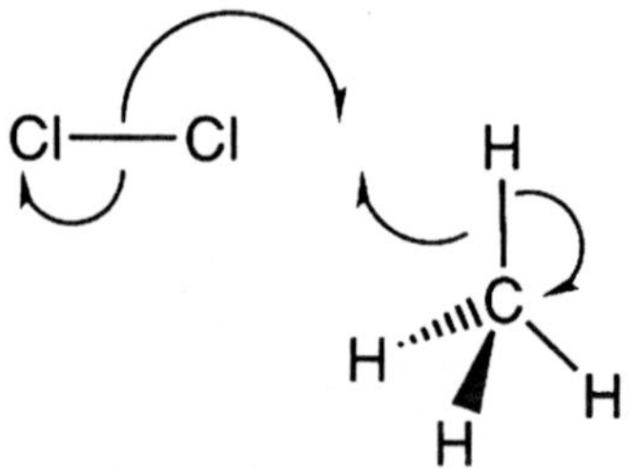

List the similarities and differences you can think of below.

In which ways are they alike?

_____

_____

_____

_____

_____

_____

_____

_____

_____

In which ways are they different?

_____

_____

_____

_____

_____

_____

_____

_____

Which of these similarities and differences do you think are important to chemists? Put a star symbol (*) in front of the important similarities, and the important differences.

Chemical comparisons

The two diagrams below show things you might study in chemistry. Think about how the things shown in the diagrams are similar and how they are different:

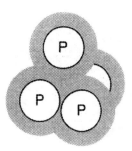

List the similarities and differences you can think of below.

In which ways are they alike?

_____

_____

_____

_____

_____

_____

_____

_____

In which ways are they different?

_____

_____

_____

_____

_____

_____

_____

Which of these similarities and differences do you think are important to chemists? Put a star symbol (*) in front of the important similarities, and the important differences.

Chemical comparisons

The two diagrams below show things you might study in chemistry. Think about how the things shown in the diagrams are similar and how they are different:

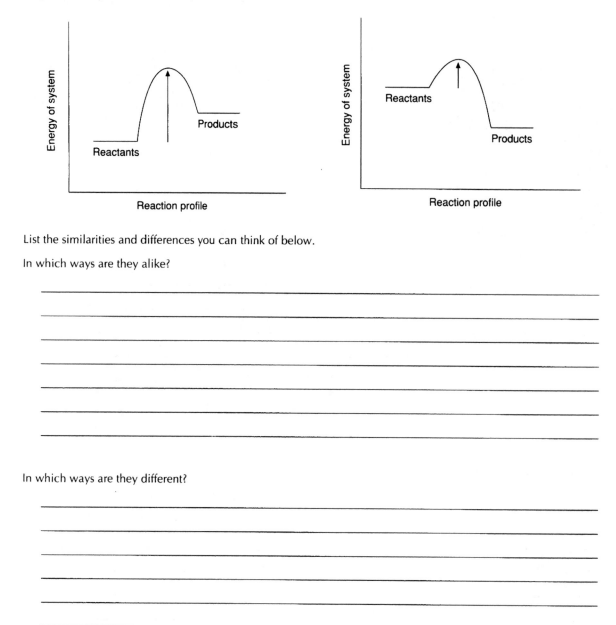

Reaction profile

Reaction profile

List the similarities and differences you can think of below.

In which ways are they alike?

_____

_____

_____

_____

_____

_____

_____

_____

In which ways are they different?

_____

_____

_____

_____

_____

_____

_____

Which of these similarities and differences do you think are important to chemists? Put a star symbol (*) in front of the important similarities, and the important differences.

Chemical comparisons

The two diagrams below show things you might study in chemistry. Think about how the things shown in the diagrams are similar and how they are different:

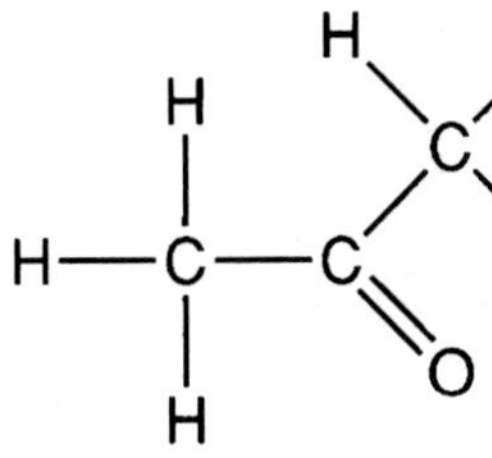

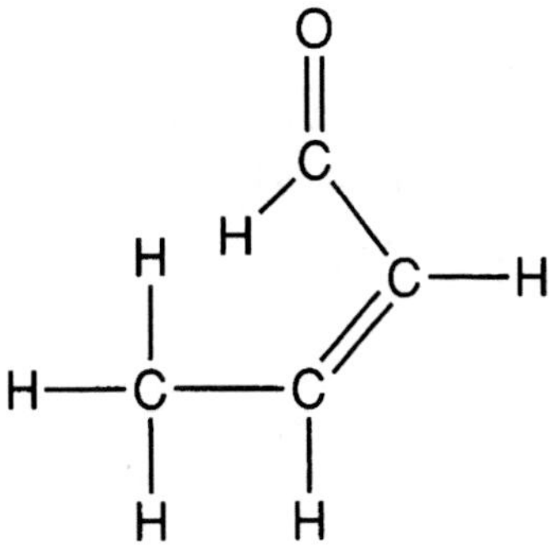

List the similarities and differences you can think of below.

In which ways are they alike?

_____

_____

_____

_____

_____

_____

_____

In which ways are they different?

_____

_____

_____

_____

_____

_____

_____

Which of these similarities and differences do you think are important to chemists? Put a star symbol (*) in front of the important similarities, and the important differences.

Chemical comparisons

The two diagrams below show things you might study in chemistry. Think about how the things shown in the diagrams are similar and how they are different:

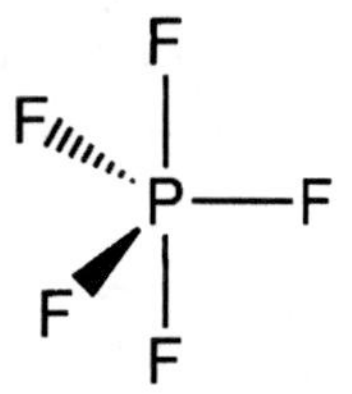

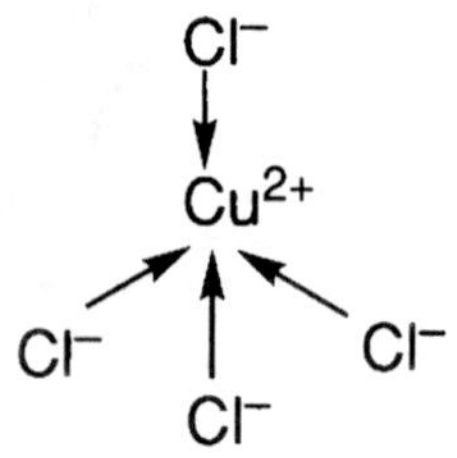

List the similarities and differences you can think of below.

In which ways are they alike?

_____

_____

_____

_____

_____

_____

_____

In which ways are they different?

_____

_____

_____

_____

_____

_____

_____

Which of these similarities and differences do you think are important to chemists? Put a star symbol (*) in front of the important similarities, and the important differences.

RS•C

Chemical comparisons

The two diagrams below show things you might study in chemistry. Think about how the things shown in the diagrams are similar and how they are different:

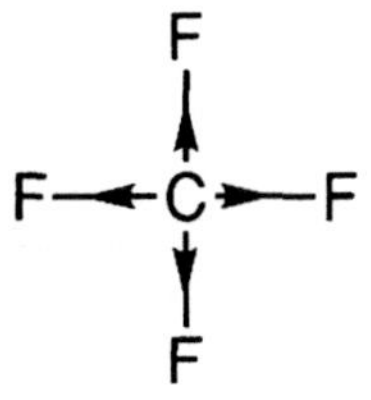

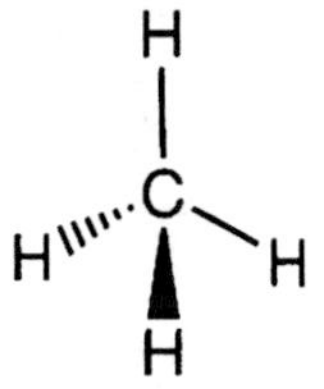

List the similarities and differences you can think of below.

In which ways are they alike?

_____

_____

_____

_____

_____

_____

_____

_____

In which ways are they different?

_____

_____

_____

_____

_____

_____

_____

Which of these similarities and differences do you think are important to chemists? Put a star symbol (*) in front of the important similarities, and the important differences.

Chemical comparisons

The two diagrams below show things you might study in chemistry. Think about how the things shown in the diagrams are similar and how they are different:

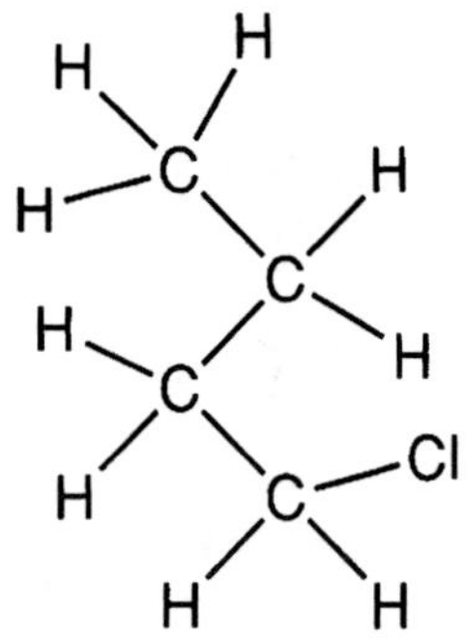

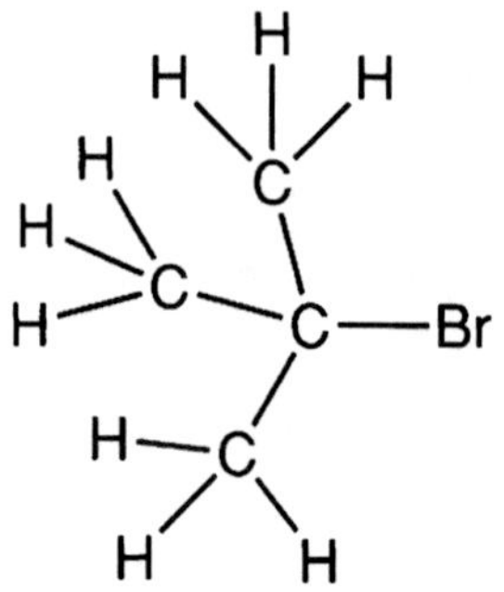

List the similarities and differences you can think of below.

In which ways are they alike?

_____

_____

_____

_____

_____

_____

_____

_____

In which ways are they different?

_____

_____

_____

_____

_____

_____

_____

_____

Which of these similarities and differences do you think are important to chemists? Put a star symbol (*) in front of the important similarities, and the important differences.

Chemical comparisons

The two diagrams below show things you might study in chemistry. Think about how the things shown in the diagrams are similar and how they are different:

List the similarities and differences you can think of below.

In which ways are they alike?

_____

_____

_____

_____

_____

_____

_____

In which ways are they different?

_____

_____

_____

_____

_____

_____

_____

Which of these similarities and differences do you think are important to chemists? Put a star symbol (*) in front of the important similarities, and the important differences.

Chemical comparisons

The two diagrams below show things you might study in chemistry. Think about how the things shown in the diagrams are similar and how they are different:

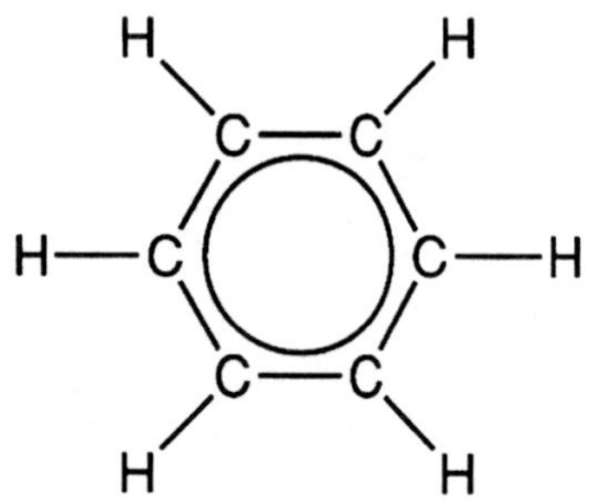

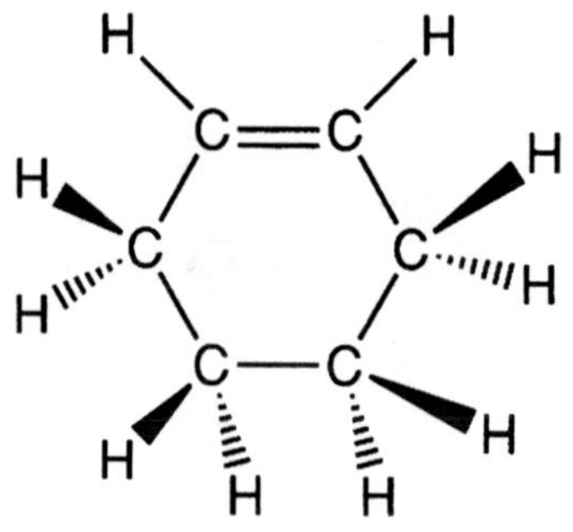

List the similarities and differences you can think of below.

In which ways are they alike?

_____
_____
_____
_____
_____
_____
_____
_____

In which ways are they different?

_____
_____
_____
_____
_____
_____
_____

Which of these similarities and differences do you think are important to chemists? Put a star symbol (*) in front of the important similarities, and the important differences.

RS•C

Chemical comparisons

The two diagrams below show things you might study in chemistry. Think about how the things shown in the diagrams are similar and how they are different:

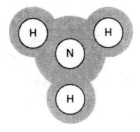

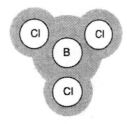

List the similarities and differences you can think of below.

In which ways are they alike?

_____

_____

_____

_____

_____

_____

_____

_____

In which ways are they different?

_____

_____

_____

_____

_____

_____

_____

Which of these similarities and differences do you think are important to chemists? Put a star symbol (*) in front of the important similarities, and the important differences.

RS•C

Chemical comparisons

The two diagrams below show things you might study in chemistry. Think about how the things shown in the diagrams are similar and how they are different:

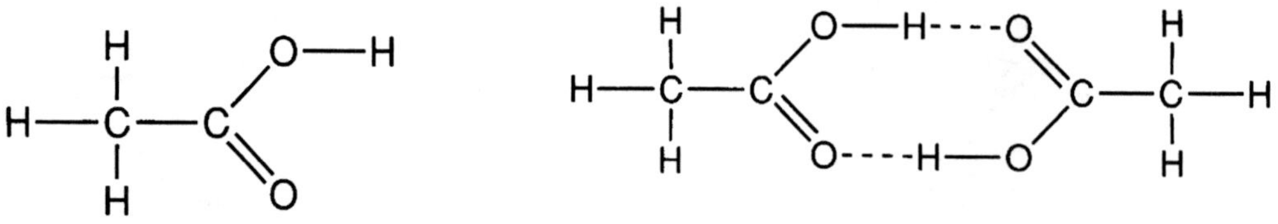

List the similarities and differences you can think of below.

In which ways are they alike?

_____

_____

_____

_____

_____

_____

_____

In which ways are they different?

_____

_____

_____

_____

_____

_____

_____

Which of these similarities and differences do you think are important to chemists? Put a star symbol (*) in front of the important similarities, and the important differences.

RS•C

Learning impediment diary

Resource

This resource is a proforma for teachers who wish to keep a record of learning impediments they meet in their teaching.

Rationale

Students' 'failures to learn' may have many causes, some of which the teacher has little influence over. Many failures, however, can be seen to at least partially due to the students prior knowledge not matching up to the teacher's expectations. This can be due to gaps in knowledge, or the presence of alternative conceptions, or simply the failure of the student to 'make the connection' that the teacher intends.

These ideas are discussed in Chapter 4 of the Teachers' notes. It is suggested there that diagnosing the source of the problems - the nature of the 'learning impediments' - can be very useful for the teacher. Such diagnosis can provide feedback useful for planning future teaching, as well as when helping the individual learner. In Chapter 4, a simple scheme for classifying 'learning impediments' is presented, accompanied by suggestions for how the different types of impediment are best tackled.

Instructions

Instances of students not understanding, or 'understanding differently' (misunderstanding), material that has been covered in class are commonly recognised in classroom exchanges, tests, homework exercises etc. A **Learning impediment record** sheet is provided for keeping notes about learning impediments identified. Chapter 4 of the Teachers' notes provides a simple 3-step guide for using the record sheets.

Learning impediment record sheet

Date: Teaching group:

Topic:

Student(s):

Description of shortcoming:

Nature of learning impediment:

Action to be taken with student/group:

Points for future planning:

Outcome:

9 780854 043811